Developing and Managing Volunteers

Independent Study 244.a

May 2010

FEMA

Page

Page

Course Overview

About This Course

This course is part of the Professional Development Series produced by the Federal Emergency Management Agency (FEMA). It is designed for emergency management professionals with current or potential responsibility for managing volunteers or volunteer programs.

This course offers training in identifying volunteer resources and recruiting, assigning, training, supervising, evaluating, and motivating volunteers. The course also focuses on coordinating with Voluntary Organizations Active in Disaster (VOADs), nongovernmental organizations (NGOs), such as church groups or food banks, professional groups, such as physicians and mental health counselors, and business and industry. It also addresses special issues, such as unaffiliated volunteers, stress management for volunteers, and legal issues, such as workers' compensation, insurance, safety and risk management, and liability.

Note: This course does not address volunteer firefighters (though technically they are emergency management volunteers) because of the vast scope of that topic. For more information about courses offered through the National Fire Academy visit the website at www.usfa.fema.gov/nfa.

FEMA's Independent Study Program

FEMA's Independent Study Program is one of the delivery channels that the Emergency Management Institute (EMI) uses to provide training to the general public and specific audiences. This course is part of FEMA's Independent Study Program. In addition to this course, the Independent Study Program includes other courses in the Professional Development Series (PDS) as well as courses in floodplain management, radiological emergency management, the role of the Emergency Manager, hazardous materials, disaster assistance, the role of the Emergency Operations Center, and an orientation to community disaster exercises.

FEMA's independent study courses are available at no charge and include a final examination. You may apply individually or through group enrollment. When enrolling for a course, you must include your name, mailing address, social security number, and the title of the course in which you wish to enroll.

FEMA's Independent Study Program (Continued)

If you need assistance with enrollment, or if you have questions about how to enroll, contact the Independent Study Program Administrative Office at:

FEMA Independent Study Program
Administrative Office
Emergency Management Institute
16825 South Seton Avenue
Emmitsburg, MD 21727
(301) 447-1200

Information about FEMA's Independent Study Program is also available on the Internet at:

http://www.fema.gov/IS

Each request will be reviewed and directed to the appropriate course manager or program office for assistance.

Course Completion

The course completion deadline for all FEMA Independent Study courses is 1 year from the date of enrollment. The date of enrollment is the date that the EMI Independent Study Office will use for completion of all required course work, including the final examination. If you do not complete this course, including the final examination, within that timeframe, your enrollment will be terminated.

Course Prerequisites

Developing and Managing Volunteers has no prerequisites.

Final Examination

This course includes a written final examination, which you must complete and return to FEMA's Independent Study Office for scoring. To obtain credit for taking this course, you must successfully complete this examination and score a 75 percent or above. You may take the final examination as many times as necessary.

When you have completed all units, you may complete the final examination online or use the answer sheet (if one is provided in your course packet). If you choose to use the answer sheet, you must return it to the FEMA Independent Study Office at the address listed on the previous page. EMI will score your test and notify you of the results.

Unit 1: Course Introduction

Introduction

The use of volunteers has proven critical to emergency management. Both individual volunteers and established volunteer groups offer a wealth of skills and resources that can be used prior to, during, and after an emergency. Mobilizing the private sector can add significantly to emergency management programs.

As an emergency management professional, your ability to work with volunteers before, during, and after an emergency can literally affect the lives and well-being of the local citizenry. Volunteers can impact—for better or worse—the ability of response agencies to do their jobs, and can make a difference in how quickly the community is able to respond to and recover from a disaster.

How to Take This Course

This independent study self-instruction course is designed so that you can complete it at your own pace. Take a break after each unit, and give yourself time to think about the material, particularly how it applies to your work as an emergency management professional, and the volunteer situations you have encountered or anticipate encountering on the job.

Developing and Managing Volunteers contains six units. Each of the units is described below.

- **Unit 1, Course Introduction,** provides an overview of the course goals and objectives, and instructions on how to take the course.

- **Unit 2, Volunteers in Emergency Management,** introduces the different types of volunteers and volunteer programs. This unit also addresses the benefits and challenges of involving volunteers, and compares creating a volunteer program to coordinating with Voluntary Organizations Active in Disaster (VOADs).

- **Unit 3, Developing a Volunteer Program,** covers the key tasks involved in working with individual volunteers and the volunteer coordinator's role in developing and maintaining an effective volunteer program.

How to Take This Course (Continued)

- **Unit 4, Working with VOADs, NGOs, and Other Groups,** examines the role of a voluntary agency/community-based organization coordinator in coordinating with various groups to identify community needs and ensure that those needs are met during an emergency. This unit covers the role that National Voluntary Organizations Active in Disaster (NVOAD) plays as an umbrella organization for coordinating with the various agencies, as well as requirements for emergency planning and coordination.

- **Unit 5, Special Issues,** points out some of the universal special issues in volunteer management, including spontaneous volunteers, volunteer stress management, and legal issues such as safety, liability, insurance, and Workers' Compensation.

- **Unit 6, Course Summary,** summarizes key concepts from the entire course.

Activities

This course will involve you actively as a learner by including activities that highlight basic concepts. It will also provide you with guidance on actions required in specific situations through the use of case studies. These activities emphasize different learning points, so be sure to complete all of them. Compare your answers to the answers provided following each activity. If your answers are correct, continue on with the material. If any of your answers are wrong, go back and review the material before continuing.

Knowledge Reviews

To help you know when to proceed to the next unit, Units 2 through 5 are followed by a Knowledge Review that asks you to answer questions that pertain to the unit content, after which, the answers are given. When you finish each Knowledge Review, check your answers, and review the parts of the text that you do not understand. Do not proceed to the next unit until you are sure that you have mastered the current unit.

Job Aids

Throughout the course, you will find job aids designed to supplement the text. You can use the job aids during the course, and you will find them useful later, after you have completed the course.

Appendixes

In addition to the six units, this course includes appendixes that contain copies of all of the job aids presented in this course, as well as an acronym list.

Final Examination

This course includes a written final examination, which you must complete and return to FEMA's Independent Study Office for scoring. To obtain credit for taking this course, you must successfully complete this examination with a score of 75 percent or above. You may take the final examination as many times as necessary.

When you have completed all units, take the final examination online or use the answer sheet (if one is provided in your course packet). EMI will score your test and notify you of the results.

Sample Learning Schedule

Complete this course at your own pace. You should be able to finish the entire course—including pretest, units, knowledge reviews, and final examination—in approximately 10 hours. The following learning schedule is only an example, intended to show relative times devoted to each unit.

Unit	Suggested Time
Unit 1: Course Introduction	$1^{1}/_{2}$ hours
Unit 2: Volunteers in Emergency Management	$1^{1}/_{2}$ hours
Unit 3: Developing a Volunteer Program	3 hours
Unit 4: Working With VOADs, NGOs, and Other Groups	2 hours
Unit 5: Special Issues	$1^{1}/_{2}$ hours
Unit 6: Course Summary	$^{1}/_{2}$ hour

Case Study: The Importance of Volunteers to Emergency Management

How critical is the ability to use volunteers in an emergency situation? The case study below shows how important volunteers can be.

Instructions: Read the following case study. As you read, think about how the organization of the volunteer management function in this community compares with that in your own community. Answer the questions that follow the case study. Then turn the page to check your answers against the answers provided.

The United Way chapter in Peculiar, MO has been approached by the interagency network of community service agencies called Peculiar Community Cares (PeCom). PeCom believes strongly that it should address disaster preparedness and response in Peculiar. The members of PeCom feel that the United Way is the ideal agency to lead this effort. The United Way chapter has a strategic goal of taking a leadership role in the community. Coordinating and developing community resources to support disaster services is, therefore, a high priority. The staff and resources in this United Way community are nearly stretched to the limit. PeCom has promised a small amount of administrative funding for the United Way if it takes on this role. Volunteerism is a strong tradition in this community, and a number of nearby colleges have offered internships in community service.

Questions:

1. What can or should the United Way do to address this community need in regard to volunteerism?

2. What would be the steps in the process of accomplishing the mission given to the United Way with regard to volunteerism?

Case Study: The Importance of Volunteers to Emergency Management (Continued)

Answers to Case Study

1. What can the United Way do to address this community need in regard to volunteerism?

The United Way can serve the important function as Volunteer Coordinator in Peculiar. As Volunteer Coordinator, the United Way can help with:

- Recruiting and maintaining a roster of volunteers.
- Assessing skills.
- Linking volunteers with other local service providers.
- Coordinating volunteer services in an emergency.

2. What would be the steps in the process of accomplishing the mission given to the United Way with regard to volunteerism?

Some steps that the United Way could take as Volunteer Coordinator include:

- Convening a meeting among service providers (i.e., VOADs, NGOs, business and industry) to determine what services they can provide, agree on roles, and identify volunteer needs.
- Developing common public awareness and recruitment materials.
- Serving as a central point for volunteer inquiries, initial interest surveys, and skills screening.
- Linking volunteers to agencies or organizations that match the volunteers' skills with the organization's needs.
- Coordinating services during and after emergencies.

Course Goals

In *Developing and Managing Volunteers*, you will learn how to work with volunteers and VOADs before, during, and after an emergency. This course will provide you with a foundation of knowledge that will enable you to:

- Describe the differences between a Volunteer Program Manager and a VOAD Coordinator.

- Determine whether you need to coordinate volunteers or work with VOADs—or both.

- Identify the skills and knowledge required of volunteers in emergency management programs.

- Develop an action plan for recruiting, interviewing, training, supervising, and evaluating volunteers.

- Develop a plan for working with VOADs, professional groups, and business and industry.

- Identify special issues in volunteer management and at least one point of contact who can provide expertise in each issue area.

Goal Setting

What do you hope to gain through completing *Developing and Managing Volunteers*? Depending on your role in emergency management, your prior experience in working with volunteers, and your current level of expertise in these areas, your goals may be slightly different from those of other emergency management professionals.

Clarifying your goals will help you gain the most from the time you spend completing this course. Take a few minutes to complete the following activity.

Activity: Developing Personal Learning Goals

Purpose:

The purpose of this activity is to help you develop personal goals for this course.

Instructions:

1. Consider the following information:

   - The course goals.
   - Your own experience with volunteers. Reflect on emergencies in which you have participated, the volunteers who worked in those emergencies, and the outcomes. What have you learned—either positive or negative—from those experiences?

2. Think about what you would like to accomplish through this course. Then list three (or more) personal goals for improving your ability to develop and manage volunteers.

<table>
<tr><td colspan="2" align="center">Goals</td></tr>
<tr><td>1.</td><td>_____</td></tr>
<tr><td>2.</td><td>_____</td></tr>
<tr><td>3.</td><td>_____</td></tr>
</table>

Unit 2: Volunteers in Emergency Management

Introduction

The United States has a long history of volunteerism. People of all ages and with all types of skills volunteer—and, with the current generation's enthusiasm for volunteerism, the current trend can be expected to continue.

So what does this mean for your volunteer program? In a nutshell, it means that your volunteer program can become as important to your community as you make it. Americans *want* to volunteer. Your job will be to develop a program that matches volunteer skills to agency needs so that both the volunteer *and* your agency accomplish their goals.

In this unit, you will learn about the different types of volunteers and volunteer programs. After completing this unit, you should be able to:

- Define volunteer and Voluntary Organizations Active in Disaster (VOAD) and draw distinctions between the two.

- List three benefits of involving volunteers in your emergency management program.

- Identify the three greatest challenges you face in developing a volunteer program.

- Determine whether your community's needs are best met by developing a volunteer program or whether you should coordinate with VOADs—or both.

Volunteers or Voluntary Organizations?

We will begin by defining terms. Although you probably have a general understanding of a volunteer as an unpaid worker, for the purposes of this course we will use the National Response Framework definition:

*A **volunteer** is any individual accepted to perform services by the lead agency (which has authority to accept volunteer services) when the individual performs services without promise, expectation, or receipt of compensation for services performed. See 16 U.S.C. 742f(c) and 29 CFR 553.101.*

Volunteers or Voluntary Organizations? (Continued)

Think for a moment about why people volunteer. Then, write several reasons for volunteering in the space below.

People volunteer for a number of reasons, including wanting to:

- Give back.

- Share their abilities.

- Develop new skills.

- See a mission accomplished.

- Network with people.

The reasons people volunteer are as broad as the type of people who volunteer. Perhaps a good place to start, then, is by looking at the types of volunteers.

Types of Volunteers

1. **Affiliated.** These volunteers are attached to a recognized voluntary organization that has trained them for disaster response and has a mechanism in place to address their use in an emergency.

2. **Unaffiliated.** Unaffiliated volunteers, also known as spontaneous volunteers, are individuals who offer to help or self-deploy to assist in emergency situations without fully coordinating their activities. They are considered "unaffiliated" in that they are not part of a disaster relief organization.

A *Voluntary Organization Active in Disaster* (or *VOAD*) is an established organization whose mission is to provide emergency services to the community through the use of trained volunteers. Examples of VOADs include The American Red Cross and many church-related agencies such as The Salvation Army, Mennonite Disaster Services, or the Southern Baptist Disaster Relief. Most, if not all, of these organizations have registered nonprofit (501(c)3) status, and many belong to the National Voluntary Organizations Active in Disaster (NVOAD). Refer to Job Aid 4.1 on page 4.5.

In addition to VOADs, there are many NGOs (e.g., Meals on Wheels, Veterans of Foreign Wars (VFW), the Elks, and church groups) whose mission is not specifically disaster-related but that, nevertheless, can be an important source of volunteer services in an emergency. Furthermore, many businesses and corporations have volunteer programs that offer goods and services that communities can tap in emergency situations. Other sources of volunteers include:

- Youth groups (e.g., from local churches, schools, or community recreation associations).

- Retired persons (e.g., the American Association of Retired Persons (AARP)).

With all of the potential sources of volunteers and volunteer services that are available, how do you decide what's best for your agency? The remainder of this unit will help you to decide.

Creating a Volunteer Program vs. Coordinating With Voluntary Organizations Active in Disaster (VOADs)

One of the first decisions volunteer managers must make is whether to develop their own agency volunteer programs or to coordinate with volunteer groups that are already established in the community.

To make this decision, it is important to consider both the positive and negative points of volunteer participation. Knowledge of the positive factors will help you enlist internal support and make your case to decisionmakers for a volunteer program. Awareness of the negative factors can help you mitigate some of them with planning and view the situation realistically. Furthermore, if the positives do not outweigh the negatives in your particular situation, the decision not to develop or continue an agency volunteer program will be obvious. So let us examine both the challenges and benefits of working with volunteers.

Involving Volunteers—Benefits and Challenges

Think about some of the benefits you've realized in working with volunteers.

Volunteers offer much more to emergency management than free labor. In fact, the benefits of involving volunteers are many. Volunteers:

- Provide services more cost effectively.

- Provide access to a broader range of expertise and experience.

- Increase paid staff members' effectiveness by enabling them to focus their efforts where they are most needed or by providing additional services.

- Provide resources for accomplishing maintenance tasks, or upgrading what would otherwise be put on the back burner while immediate needs demand attention.

- Enable the agency to launch programs in areas in which paid staff lacks expertise.

- Act as liaisons with the community to gain support for programs.

- Provide a direct line to private resources in the community.

- Facilitate networking.

- Increase public awareness and program visibility.

Involving Volunteers—Benefits and Challenges (Continued)

Along with the many benefits of involving volunteers, there are also challenges—some real and some perceived—involved with working with volunteers.

- Training and supervision of volunteers take a lot of time.

- Volunteers do not stay, so the time spent training them is wasted.

- Technically competent people do not volunteer.

- Volunteers threaten paid staff by competing with them.

- Volunteers lower professional standards.

- Volunteers become territorial and/or attempt to take over.

- Insurance rates will increase.

- Volunteers are not available during business hours.

- Using volunteers interferes with the ability to negotiate for additional funding or new paid staff positions.

Note that training and supervision is an actual challenge inherent with both volunteers and paid staff.

Many of these challenges often boil down to misperceptions, or are a result of poor planning and management.

Common complaints of both volunteers and paid staff are listed on the next page. As you read them, think about how many of them represent genuine challenges and how many are the result of poor management.

Involving Volunteers—Benefits and Challenges (Continued)

Common Volunteer Complaints	Common Staff Complaints
• I'm always told what to do but never asked to participate in planning the work. • Salaried staff gets or takes the credit for my good ideas. • No one says, "Thank you." • I always seem to get the "grunt work." • I never get feedback on my work. • Salaried staff is always given the benefit of the doubt in any dispute. • Can't I have a title other than "Volunteer?" • I always have to search for a place to do my work. There is nowhere to store my work from week to week.	• Volunteers will take salaried jobs. • Volunteers will do a poor job and I'll get the blame and have responsibility for cleaning it up. • Volunteers will do a great job and I'll look less effective. • Volunteers are amateurs. • Volunteers gossip; they don't understand confidentiality. • Volunteers are not dependable. • Volunteers will lower professional standards. • Volunteers get all the recognition.

As you review both volunteer and staff complaints, you should note that all can be mitigated by effective planning and management. For example, one common theme is an "us vs. them" mentality between staff and volunteers. Knowing that this potential exists enables you to address this issue actively by:

- Treating volunteers with respect. Remember that people volunteer for a number of reasons, but none of them involve being treated poorly. Recognize all that volunteers can offer, and treat them with the respect they deserve.

- Conducting teambuilding activities with volunteers and staff. All organizations work better when their personnel work together. Conducting simple activities and discussions to air differences and emphasize the importance of working toward a common goal will benefit both the staff and volunteers—and improve the overall emergency management function.

- Reducing the perceived threat to staff members by educating them about the benefits that volunteers provide to them as well as to the agency. If managed well, volunteers will make the paid staff's jobs easier by allowing them to focus on the parts of their jobs that only they can do.

Involving Volunteers—Benefits and Challenges (Continued)

Some of the benefits to working with volunteers as a staff member are listed below:

- Supervising volunteers can demonstrate managerial ability.

- Supervising volunteers can provide on-the-job experience for employees seeking promotion to managerial positions.

- Volunteers can reduce the overall workload for everyone.

- Volunteers can do tasks that staff does not have time to do.

The checklist on the next page will help you determine your organization's volunteer-staff climate.

Job Aid 2.1: Checklist for Determining Volunteer-Staff Climate

Instructions: *Review each of the statements listed below and mark those that you think accurately reflect the climate in your organization. When you finish, review the list. If only a few boxes are checked, you have some work to do to develop a healthy volunteer program.*

☐   Our organization is stable and conflict-free, with a healthy work environment.

☐   Agency policy places high priority on volunteer involvement.

☐   We have established clear, realistic goals for volunteer involvement.

☐   Staff and volunteer roles and responsibilities have been clearly defined and documented.

☐   Volunteer job descriptions were developed with input from staff members and take into account work assistance needs.

☐   The agency recruits and hires staff members who are experienced with and enthusiastic toward working with volunteers.

☐   Our agency ensures that volunteers work primarily with staff members who are receptive to volunteers.

☐   We have established a training program to ensure volunteers and staff members will work together effectively.

☐   Volunteer orientation includes training on sensitivity for staff problems.

☐   We have established a system for rewarding and recognizing volunteers as well as their staff supervisors.

Involving Volunteers—Benefits and Challenges (Continued)

Many perceived obstacles about using volunteers are actually misperceptions and the result of poor planning and/or management. You may face very real challenges when developing a volunteer program in your agency, such as a sparse population spread over great distances (in rural areas) from which to draw volunteers. Even these challenges can often be overcome through adequate planning, good management, and some creativity (for example, through the use of technology in rural areas).

Example:

Read the scenario below and decide whether it is illustrative of an obstacle to involving volunteers.

Scenario:

You have received reports that one of your volunteers is becoming confrontational with victims coming in to inquire about disaster assistance. After speaking with the volunteer, you learn that she has lost a family member in the disaster.

How would you handle this situation?

Scenario Answer:

This scenario does not involve an obstacle but a misperception. Managing all personnel—whether paid or volunteer—entails the potential for personal issues. This reality is not a challenge to involving volunteers per se but a corollary of working with people in general.

Probably the best way to handle this volunteer is to refer her to a counselor and reassign her to a job in which she does not have to deal directly with victims.

Volunteers? Or Voluntary Organizations?

In some cases, the challenges of involving volunteers may, in fact, outweigh the benefits. It may be better to work through VOADs—at least until you can work through the challenges and create a better program that will meet the needs of both the agency and the volunteer. Be sure you understand the responsibilities, mission, and capabilities of the voluntary agencies you may want to access. Establishing good working relationships with the VOADs will enable you to understand which agency can assist you, as well as agency needs, capabilities, and limitations.

The activity that follows asks you to think about the benefits and challenges in your specific situation of developing or maintaining your own volunteer program vs. coordinating with VOADs.

Activity: Benefits and Challenges of Using Volunteers

Use the questions below to compare the relative advantages and disadvantages of developing and/or maintaining your own agency volunteer program versus coordinating with VOADs and NGOs in your community for volunteer services.

1. How have you used volunteers in the past?

2. How successful have volunteer efforts been?

3. How willing are individuals to volunteer in the community?

4. What VOADs and NGOs are present in your community that you could coordinate with for needed volunteer services?

5. What are three benefits of having your own volunteers in your particular agency situation?

6. In the left column below, list the three greatest challenges you can foresee to developing (or maintaining, if you already have one) your own agency volunteer program. Then, in the right column, write the best resource(s) to overcome each challenge.

Challenge	Resource(s)
Example: Training	*VOAD training manuals/programs*
1.	
2.	
3.	

Unit Summary

This unit provided an overview of the benefits and challenges of involving volunteers in emergency management. It also asked you to begin considering whether developing and/or maintaining your own agency volunteer program or coordinating with existing VOADs would be best for your situation. To help you further explore that question, the next unit examines what is involved in developing and maintaining a volunteer program.

For More Information

Volunteer Management Links

- National Voluntary Organizations Active in Disaster

  http://www.nvoad.org/

- Volunteer Management/Service Leader Resources

  http://www.serviceleader.org/

- Citizen Corps

  http://www.citizencorps.gov/

Knowledge Review

Select the best answer. Turn the page to check your answers.

1. The main benefit of involving volunteers is that they provide free labor.

   a. True
   b. False

2. Tension between paid staff and volunteers is inevitable.

   a. True
   b. False

3. Most perceived challenges to the use of volunteers are actually:

   a. Accurate.
   b. Misperceptions.
   c. Poor management.
   d. Matching volunteers to the wrong job.

4. One benefit of involving volunteers for staff members is that volunteers can be assigned all of the unpopular jobs.

   a. True
   b. False

5. One strategy for improving the staff-volunteer working climate is to treat both staff and volunteers with equal respect and consideration.

   a. True
   b. False

Knowledge Review (Continued)

1. b
2. b
3. b
4. b
5. a

Unit 3: Developing a Volunteer Program

Introduction

In this unit, you will learn about the key tasks required to develop a volunteer program and work with individual volunteers. After completing this unit, you should be able to:

- Describe the roles and responsibilities of a Volunteer Program Director.

- Determine how volunteers can be used most beneficially in your program to meet your agency's needs.

- Design a volunteer program.

- Write a volunteer job description.

- Develop a strategy for recruiting, assigning, training, supervising, and evaluating volunteers.

The Roles and Responsibilities of a Volunteer Program Director

If your agency chooses to develop and/or manage its own agency volunteer program, it will no doubt need one person to be in charge of the program (referred to here as the Volunteer Program Director). Whether that person is you or someone that you coordinate with, the Volunteer Program Director's responsibilities should include:

- Planning for volunteer involvement.

- Overseeing the implementation of the overall volunteer strategy.

- Setting volunteer program policy, perhaps in conjunction with a committee.

- Developing and managing the volunteer budget. (Don't forget the cost of developing training manuals, Standard Operating Procedures (SOPs), etc.)

- Promoting and publicizing the volunteer program.

The Roles and Responsibilities of a Volunteer Program Director (Continued)

- Recruiting, selecting, assigning, training, and supervising the volunteers.

- Coordinating with staff and programs with which the volunteers interface.

Each of these tasks should be developed as part of the process of volunteer program design.

One way to ensure that you have the most professional volunteer program possible is to encourage your Volunteer Program Director or Voluntary Agency Coordinator (to be discussed in Unit 4) to get certified by the Council for Certification in Volunteer Administration (CCVA). CCVA is the international professional association that awards the Certified in Volunteer Administration (CVA) credential. For further information contact:

Council for Certification in Volunteer Administration
P.O. Box 467
Midlothian, VA 23113
CCVA@comcast.net
(804) 794-8689

Volunteer Program Structure

Before you start working with individual volunteers, you must step back and look at the big picture. In other words, consider the mission of your agency and look at what tasks are not getting done because of staffing shortfalls. You should then do a needs analysis to identify possible volunteer roles to fill tasks to be done; establish job descriptions; recruit, place, train, supervise and evaluate the individual volunteers; and evaluate your volunteer program.

You should plan on a *minimum* of 6 months when developing a volunteer program from the beginning. Depending on how many volunteers will be recruited, the skills they bring to the job, and the amount of training they require, it could take much longer.

Steps in Developing a Volunteer Program

There are seven steps in developing your volunteer program, including several that concern working with individual volunteers. One goal of performing the activities in the steps is to get you started on creating products tailored to your own use when you return to your workplace and work with volunteers.

The steps are:

1. Analyzing Agency and Program Needs

2. Writing Volunteer Job Descriptions

3. Recruiting Volunteers

4. Placing Volunteers

5. Training Volunteers

6. Supervising and Evaluating Volunteers

7. Evaluating Program

Step 1: Needs Analysis

Before beginning the process of developing a volunteer program, you must determine the needs of your agency or organization. In other words, what functions do you need volunteers to perform within your agency?

Defining agency needs involves:

- **Considering the agency's mission.** What is the agency's mission? How well does it meet that mission? (This step may require completing a community emergency management needs analysis, especially if your agency does not have a well-defined mission statement.)

- **Looking at current staffing resources and areas of shortfalls where volunteers may be able to help.** What tasks are consistently relegated to the back burner because there is never enough time or personnel to do them? What good ideas are not being implemented for the same reasons? What portions of the agency's mission could be accomplished better if more resources were available?

- **Describing:**

  - **Tasks** that need to be done.
  - The **skills, knowledge, and abilities** necessary to do the tasks.
  - Additional **resources** (e.g., equipment, office space) needed.

Job Aid 3.1 includes a checklist of Basic Criteria for Developing a Volunteer Job. Use this checklist as a guide to determining whether functions or positions could be filled by volunteers.

When designing jobs for volunteers, keep in mind that the jobs should reflect not only the needs of your agency but of volunteers as well. Volunteers' needs include the desire to contribute and make a difference, social needs to interact with other people, an interest in (and the ability to perform) the job itself, and the desire to learn new skills. Consider how marketable the potential job is to volunteers. Making the hours flexible and including resume-building opportunities are two ways to make a job more attractive to potential volunteers.

Step 1: Needs Analysis (Continued)

A well-defined job description is also a valuable tool for the Volunteer Program Director. A good job description is:

- The first step in the recruitment process. Identifying and defining needs helps to target volunteers to fill those needs.

- A tool for marketing the agency's need to potential volunteers.

- A focal point for interviewing, screening, and selection.

- The basis for performance evaluation.

When planning for volunteer involvement, don't forget to identify and address concerns that paid staff may have about working with volunteers. Involve them in the development of the volunteer program through such tasks as:

- Writing volunteer program policy and job descriptions.

- Volunteer training.

Involving paid staff gives them a stake in making the volunteer program succeed.

Job Aid 3.1: Basic Criteria for Developing a Volunteer Job

☐ Is there meaningful work for volunteers to do? (Consider its significance to the agency and how to explain the need for the job to potential volunteers.)

Can the work be done by volunteers? (Consider ability to split tasks into part-time work, skill requirements for the job, and whether tasks would be short- or long-term assignments.)

☐ Is it cost-effective to use volunteers? (Consider time, energy, and money for recruitment, orientation, and training of volunteers.)

☐ Is a support framework in place for a volunteer program, including:

- A volunteer manager?
- A volunteer policy?
- Volunteer work space?
- Insurance covering volunteers?

☐ How will paid staff work with volunteers? (Consider experience and receptiveness, as well as role and responsibility definitions for staff and volunteers.)

☐ How will you find volunteers with skills to do the job? (Consider recruitment tactics as well as orientation and training programs.)

☐ Has the agency committed to volunteer involvement with clear policy?

Activity: Volunteers and Incident Management Activities

This activity will provide an opportunity for you to think about the roles that volunteers could play in each of the incident management functions. Complete the matrix below. One example in each function has been suggested to get you started.

Function	Potential Volunteer Roles
Prevention	*Organize a local Neighborhood Watch program.*
Preparedness	*Organize local disaster preparedness drills.*
Mitigation	*Develop materials to educate the public to potential mitigation methods or present mitigation suggestions—protective actions such as anchoring water heaters and shelves to a community that is at risk from an earthquake.*
Response	*Answer emergency helplines.*
Recovery	*Assist disaster victims with filling out insurance forms and/or registering for State or Federal aid.*

Step 2: Writing Volunteer Job Descriptions

When you have completed your needs analysis, your next task is to develop a job description for each position that you have identified. Because a job description may be used as a legal document, it should be as complete as possible—and you should have it reviewed by legal counsel before you begin recruiting volunteers for the position.

When developing a job description, think about:

- The purpose of the job. How will the position help your agency achieve its mission?

- The job responsibilities. What primary tasks will you expect the volunteer to do? What occasional tasks will you expect?

- Job qualifications. What knowledge, skills, and abilities (KSAs) are required for the job? Should the volunteer have attained a certain education level to be successful?

- To whom the volunteer will report. Will he or she report directly to you? To a member of your staff? To another volunteer?

- The time commitment required for the position. How many hours each week are required to ensure that the job responsibilities can be accomplished within a reasonable timeframe and without undue stress?

- The length of the appointment. How long will the position be required? Is the job open-ended, or is it a position that is only required during an emergency?

- Who will provide support for the position. Will the volunteer work independently, or will he or she rely on others in the organization (e.g., publications, outreach)?

- Development opportunities. Is formal training available or required for the position? What are the possibilities for advancement?

Considering each of these factors will help you develop a solid job description and make developing a recruitment strategy easier.

A sample format for a job description is shown as part of the activity on the next page.

Activity: Writing a Volunteer Job Description

This activity will provide an opportunity for you to write a volunteer job description. Select one of the volunteer roles you identified in the previous activity. Then, fill in the sample form below to define the job.

Job Title   *List the volunteer job you're describing.*

Purpose   *Describe the purpose of the position in terms of benefits to your organization and to the community.*

Responsibilities   *List major duties.*

Qualifications   *List education, experience, knowledge, and/or skills required.*

Supervisor   *Be sure the volunteer is aware of who he/she reports to.*

Relationships   *Who will the volunteer work with or receive support from?*

Time   *Include the number of hours per week required, and state if flexible. Specify if the appointment open-ended or of limited duration.*

Support   *List equipment provided, travel reimbursement, training, workspace, etc. Also describe opportunities for development, such as conferences and seminars.*

Step 3: Recruiting Volunteers

The critical step in getting your program started is getting the right volunteers to fill the jobs.

Recruitment can be broad-based (general) or targeted (selective). Use a broad-based recruitment effort when there is a need for a large number of individuals for jobs that require commonly possessed skills—for example, if you need many volunteers in a potential flood emergency to place sandbags on the banks of a rapidly rising river. You can also use general recruitment when you need a wide variety of skills. Broad-based appeals for volunteers can be made to the general public through mass media outlets. The President's call for volunteers to join the Freedom Corps is a good example of broad-based recruitment.

Use targeted recruitment when you are looking for volunteers with specific skills to do specific jobs. In the case of targeted recruitment, an appeal is usually made to a targeted group through specialized media, such as association newsletters, or through personal contact. Targeted recruitment is usually more effective than general recruitment. A good example of targeted recruitment is medical research facilities that recruit volunteers who have specific health conditions for drug-trials.

Determining whether recruitment should be broad-based, targeted, or a combination of both will help you develop a recruitment strategy.

Developing a Recruitment Strategy

The next task in recruitment is to investigate the marketplace of potential volunteers and develop a recruitment strategy. Some ways to develop a recruitment strategy include:

*   Developing a short list based on what you know about the volunteers you are trying to reach.

*   Determining ways to reach those volunteers. Some possibilities include:

    *   Networking with businesses and community groups.
    *   Identifying organizations that have an impact on your agency, or, conversely, groups that your agency impacts. Determining who has what you need will enhance your recruitment efforts.
    *   Analyzing what is being done by other agencies to recruit effectively and identifying ways to make your agency more attractive to volunteers.

Activity: Listing Recruitment Sources

In the space below, jot down some ideas of groups that you might network with and places to recruit.

Source	Potential Roles

Activity: Listing Recruitment Sources (Continued)

Some potential sources for volunteers are shown below. The roles that these volunteers could fill are very broad.

Source
• High schools
• Colleges and Universities
• Alumni associations
• Businesses
• Churches
• Civic groups
• Professional organizations
• Service clubs
• Fraternal clubs
• Social service agencies
• Trade associations
• AARP (or other senior citizen groups)
• YMCA and YWCA
• Community or neighborhood groups
• Parent-Teacher Associations (PTA)
• Tenant councils
• Toastmasters
• Volunteer centers

Spend time thinking about the types of volunteers you need and how those volunteers may be reached. Spending time up front can save both time and money later.

Activity: Developing a Recruitment Strategy (Continued)

Finally, select one or more media that are likely to reach the volunteers you want. When selecting media:

- Be sure to match the media outlets to the probable demographics of your targeted group. For example, while some senior citizens may listen to heavy-metal radio stations, most probably don't. It would be more effective to recruit senior citizens via newspapers or senior citizens groups.

- Select the media types that are within your budget. Television time is more expensive than radio time, and both are more expensive than newspaper space. If other factors are equal, why pay more?

Careful selection of the media you intend to use can enhance your opportunities for reaching the audience you want.

For more information about selecting media, refer to the *Effective Communication* self-instruction course (IS 242), which is available at http://training.fema.gov/IS.

Activity: Developing a Recruitment Plan

This activity will provide an opportunity for you to develop a recruitment plan. Using the same volunteer job, write a recruitment plan in the space provided below. Element prompts are included to guide you.

Volunteer Position   *[Assume use of job description from activity on writing a job description]*

Targeted Audience

Medium

Developing a Recruitment Message

After you have determined who and where to recruit, the next task is to write a message that will reach your potential volunteers, keeping in mind what needs of volunteers you are appealing to. Also keep in mind what medium you are writing for, as the medium will affect how you write the message (e.g., the length, whether you need to include a visual element, etc.).

Regardless of the medium selected, the message should include these elements:

- An opening that will catch the audience's attention.

- A statement of the need: What is the problem?

- A statement of the solution: How volunteers can meet the need.

- A statement to address the listener's question as to whether he or she can potentially do this job.

- What's in it for the volunteer?

- A contact point to get involved.

Developing a Recruitment Message (Continued)

Read the following recruitment message, and think about whether it would be effective based on the criteria listed above. Also, think about what needs of volunteers it might be appealing to.

In 95 percent of all emergencies, bystanders or victims themselves are the first to provide emergency assistance or to perform a rescue. It starts with you.

Citizen Corps was created to capture the spirit of service throughout our communities and to help answer two key questions being asked by citizens, "What can I do?" and "How can I help?" No matter where you live, no matter who you are, we all have a role in hometown security.

Citizen Corps asks you to embrace the personal responsibility to be prepared; to get training in first aid and emergency skills; and to volunteer to support local emergency responders, disaster relief, and community safety.

Citizen Corps was created to help coordinate volunteer activities that will make our communities safer, stronger, and better prepared to respond to any emergency situation. It provides opportunities for people to participate in a range of measures to make their families, their homes, and their communities safer from the threats of crime, terrorism, and disasters of all kinds.

To find the council nearest you, go to www.citizencorps.gov.

Notice that the above message contains an attention-grabbing opening (*"In 95 percent of all emergencies, bystanders or victims themselves are the first to provide emergency assistance…"*). It defines the need (personal preparedness, training, and volunteerism). It specifies the solution (Citizen Corps councils). It holds out the possibility that the listener can do the job (*"…we all have a role in hometown security."*) In addressing benefits to the audience, it appeals to potential volunteers' need to make a difference (*"…two key questions being asked by citizens, 'What can I do?' and 'How can I help?'"*). Finally, it specifies a contact (*"go to www.citizencorps.gov"*).

In the two activities that follow, you will develop a recruitment plan, then write your own recruitment message.

Activity: Writing a Recruitment Message

This activity will provide an opportunity for you to write a recruitment message. Use the volunteer job description you wrote in the previous activity and write a recruitment message in the space below. Prompts for the necessary elements are included to guide you.

"Catchy" Opening

Need

Solution

Address listener's question about his/her ability to do the job

Benefits

Contact Point

Step 4: Placing Volunteers

Placing volunteer applicants into the appropriate job involves two processes:

- Screening

- Interviewing

Screening

All volunteers should receive some level of screening to ensure that they are a good enough fit for the job to warrant an interview. The intensity of the screening is determined by the amount of risk associated with the position for which the volunteer is applying. Job risk factors include:

- Amount of time the volunteer will spend unsupervised.

- Work with vulnerable populations (e.g., children, the elderly, people with disabilities).

- Requirement to handle funds.

- Requirement to operate a vehicle.

- Level of physical risk to the volunteer.

High-risk jobs involving working with vulnerable populations or handling money generally require a criminal background check as part of the screening process.

Common screening tools include:

- **The application.** The application is the initial screening tool because it asks for information about the volunteer's background that can be used to determine whether the applicant has the necessary experience and/or skills to be matched to the available job. The application also provides basic information and should include a consent form, both of which are necessary for more intensive screening. The consent form should ask the applicant to:

  - Verify that the information provided is accurate.
  - Waive the right to confidentiality for screening-related purposes.
  - Consent to the types of screenings to be conducted (e.g., reference check, criminal records check, etc.).

Job Aid 3.2 on page 3.22 presents a sample application.

Screening (Continued)

- **A reference check.** Although reference checks should never be the main screening tool, they can be very useful. Questions to ask references include:

  - How long and in what capacity have you known the applicant?
  - How would you describe the applicant?
  - How does the applicant relate to coworkers, youth, and people in authority?
  - How does the applicant manage stress and/or conflict?
  - Is there anything else you would like to tell me about the applicant?
  - If you had the opportunity to work with the applicant again, would you?

It is important to attempt to speak directly to the reference contact. Assure him or her that the applicant has signed a release for this information.

- **Professional license, criminal background and child abuse clearance checks.** These types of checks are frequently mandated by Federal, State, or local laws. However, even if not required, conducting these checks is prudent, especially for high-risk positions. Be sure to verify that motor vehicle and professional licenses are current and valid in the jurisdiction in which the services are to be rendered. For positions involving work with youth, obtain a child abuse clearance. For positions involving funds management, obtain a criminal background check. Be aware, however, that these checks have limitations. For example:

  - Name searches will not find aliases.
  - Out-of-State records may not be included.
  - Some departments may not be permitted to release information.
  - Only convictions are recorded, not charges, which may have been dropped, regardless of guilt.

For potential volunteers who pass all screening requirements, the next step is the interview.

Interviewing

The interview is also a screening tool—a two-way tool. Often, the applicant is still deciding whether or not to volunteer when he or she is interviewed. For this reason, it is important to ensure that your agency makes a good first impression. The interview should take place as soon as possible after the potential volunteer applies (but after initial screening is complete) so as not to leave him or her dangling. If possible, a satisfied volunteer should participate in the interview.

When interviewing volunteers, the interviewer is marketing the organization as well as determining whether the applicant is a good fit for the job. The interviewer, therefore, should be equipped with some basic tools to conduct an effective interview:

- The potential volunteer's application.

- A form for recording the interview. Job Aid 3.3 on page 3.24 presents a sample form.

- A list of open-ended questions to ask the applicant. (Possible questions are included on the sample interview recording form.)

- Information about the agency and about current volunteer opportunities.

One good approach to conducting an interview is to compose a panel of three informed stakeholders rather than just a single interviewer. An opening question might be "What brings you to apply for this job?"

The goal of the interview should be to determine the applicant's skills and motivation for volunteering and to match the applicant with a position, if desirable. Other goals include:

- Answering the applicant's questions.

- Identifying undesirable candidates. Careful attention at this point may eliminate problems later.

Interviewing (Continued)

It is important to be aware that there are a number of questions that legally may *not* be asked during an interview. These include questions about:

- Race, national origin, or birthplace.

- Marital status.

- Religious affiliation.

- Credit card or home ownership.

- Age, height, or weight.

- Pregnancy or childcare arrangements.

- Arrest record (but criminal background checks are permissible).

- Discharge from military service.

- Length of residency in the community.

- Health. The exception is a specific question about whether the applicant is able to perform a specific physical task required by the job (e.g., lift 50 lbs.).

In general, do not ask anything that is not directly related to the ability of the applicant to perform the specific volunteer job.

The interview should result in a recommendation for further action—either for placement, additional screening, or rejection if it is obvious that your agency cannot benefit from the applicant's services, or if concerns arise that could be detrimental to the agency.

If the applicant is rejected, do not refer him or her to another agency without contacting that agency first.

Job Aid 3.2: Sample Application

Name: _____

Address: _____

City: _____ State: _____ Zip: _____

Phone: (Home) _____ (Work) _____

Contact in an emergency: _____ Phone: _____

I. Skills and Interests

Education: Degree _____ Institution _____ Dates attended _____

License(s) held: _____ Language(s) spoken fluently: _____

Hobbies, skills, and interests: _____

Occupation: _____ Employer: _____

Address: _____ Phone: _____

II. Experience (paid and volunteer, beginning with the most recent):

Position   Organization   Dates

_____

_____

_____

_____

_____

III. Volunteering Preferences

Is there a particular type of volunteer work in which you are interested? [A checklist of options can be included here.]

_____

Availability (days and hours): _____

Do you have access to a vehicle that you can use for volunteer work? ___ Yes ___ No

How did you hear about our agency? _____

Job Aid 3.2: Sample Application (Continued)

IV. References

Give the names and contact information for three people (not relatives) who know you well and can attest to your character.

_____

_____

_____

V. Verification and Consent for Reference and Background Check

I verify that the above information is accurate to the best of my knowledge.

I give [name of agency] permission to inquire into my educational background, references, licenses, police records, and employment and/or volunteer history. I also give permission to the holder of any such information to release it to [name of agency].

I hold [name of agency] harmless of any liability, criminal or civil, that may arise as a result of the release of this information about me. I also hold harmless any individual or organization that provides information to the above-named agency. I understand that [name of agency] will use this information only as part of its verification of my volunteer application.

_____   _____
Name (please print)   Social Security Number

_____   _____
Signature   Date

_____   _____
Witness   Date

Job Aid 3.3: Volunteer Interview Record

Name of Volunteer_____

Interviewer_____ Date_____

I. Review of Application Form

II. Questions

1. Why do you want to volunteer with our agency? What do you hope to achieve?

3. What kind of work do you most enjoy? Least?

4. Do you work best alone or with others? Why?

5. What kind of supervision do you prefer?

6. What questions do you have?

III. Match with Volunteer Positions

Discuss potential volunteer positions and check match of interest, qualifications, and availability. Ask if there are any physical limitations.

Job Aid 3.3: Volunteer Interview Record (Continued)

To be completed after the interview:

IV. Interviewer Assessment

Appearance:

Disposition/Interpersonal skills:

Reactions to questions:

Physical restrictions:

V. Recommended Action

___ Place as _____

___ Consider/Hold in reserve for the position of _____

___ Investigate further

___ Refer to _____

___ Not suitable for agency at this time

VI. Notification

Volunteer notified of agency decision by (method) _____ on (date)_____

Activity: Developing Interview Questions

This activity will provide an opportunity for you to develop questions to ask of potential applicants. For the volunteer position you selected in the previous activity, write five appropriate interview questions in the space provided below.

Step 5: Training Volunteers

After interviews are complete and volunteers are brought on board, the next step in their development involves training.

There are important correlations between training and volunteer satisfaction and effectiveness. Thus, a well-planned training program is critical to volunteer performance and retention.   Proper training can also:

- Save lives.

- Protect property.

- Reduce suffering.

- Diminish vulnerability to lawsuits.

Orientation

Training the volunteers you have placed begins with orientation. There are two levels of orientation:

- Agency orientation.

- Job-specific orientation.

Components of a successful orientation to the agency include:

- The mission of the agency and the agency's relationship to the community and other community agencies.

- The agency's values.

- Agency procedures and issues.

- The role of volunteers in the agency.

These components can be addressed through a group presentation of a videotape and/or various speakers, including a person who holds a senior position in your agency (for the agency's mission), outside experts (for special issues, such as liability), and current volunteers (for the role of volunteers). Agency procedures can also be addressed by giving each volunteer a policy and procedures manual.

Orientation (Continued)

(Note: Agency policies and procedures should be discussed, even if a policy and procedures manual is distributed. Discussion of key policies and procedures eliminates uncertainties about what is important and why and allows the new volunteers to ask questions or get clarification on policies and procedures they do not understand.)

Consider making the orientation part of a fun activity, such as a potluck dinner for new volunteers. Such an event sets a welcoming tone and gives the new volunteers a social opportunity to meet staff and other volunteers informally.

Job Aid 3.4 on the following page is an Orientation Checklist that you can use in planning a volunteer orientation.

Job Aid 3.4: Orientation Checklist

Before the volunteer(s) arrive(s):

___ Prepare paid staff.

___ Assign a one-on-one mentor.

___ Set up the video presentation and/or confirm date and time with speakers.

___ Collect necessary items (handbook or manual, I.D. tags, etc.).

On arrival:

___ Welcome the volunteer(s).

___ Introduce the volunteer(s) to the staff (paid and volunteer).

___ Review administrative details (phones, parking, restrooms, breaks and lunch, check in/out procedures, etc.).

___ *Optional:* Give a tour of the facility.

Materials you should give volunteers:

___ Mission statement

___ Summary of goals and/or long-range plan

___ Organizational chart

___ Policies and procedures (including emergency procedures)

___ Confidentiality policy

___ *Optional:* Map of facility

What you should tell volunteer(s) about your agency:

___ Mission and goals

___ Background and history

___ Organizational structure

___ Funding base

___ The role of volunteers in the agency

___ The agency's role in the community

___ How the agency relates to other community organizations

Orientation (Continued)

All volunteers should also receive an orientation to the job. Orientation to the specific job should include:

- Specific job responsibilities.

- Who the volunteer's immediate supervisor is and his or her expectations in terms of schedule, reporting back, and/or record keeping.

- Authority and accountability (i.e., what the volunteer can and cannot do without explicit direction).

- Other team members' roles.

- Resources available to do the job.

Orientation to the job is best accomplished by one-on-one mentoring. If possible, the mentor should be an experienced volunteer.

Training

Training, unlike orientation, does more than give information. It teaches or upgrades skills, providing the "how to" that a person needs to do his or her job.

Like orientation, training can be general or specific. General training includes training in skills that are needed in many positions, such as:

- Communication skills.

- Interpersonal skills.

- Team-building skills.

- Problemsolving and decisionmaking skills.

- Leadership and supervision skills.

- Stress management.

FEMA has developed training to address many general skill areas. Many of these courses are available in both independent study and classroom versions. (See the end of this unit for more information.)

Training (Continued)

Job-specific training teaches the skills required for the volunteer to do his or her job, such as severe weather spotting. Be aware that some jurisdictions may mandate specific training requirements, as well as refresher training for skills such as Cardio Pulmonary Resuscitation (CPR). (The American Red Cross is a good resource for CPR and first aid training.)

Training can and should be ongoing (i.e., refresher skills training). Training should be worked into a long-range schedule of agency activities and mandated for both paid and volunteer agency staff. It is important to keep good records of who completed what training and when. One way to keep volunteers' training current is to certify those who have completed training and issue cards with expiration dates.

Activity: Developing a Training Plan

This activity will provide an opportunity for you to develop a training plan. For the volunteer position you selected in the previous activity, write a rudimentary training plan in the space provided below. Element prompts are included to help you.

Sample Training Plan		
Position:		
General Training:		
Course	Timeframe for Completion	Date Completed
Job-Specific Training:		

Training (Continued)

It is a good idea to maintain a record of each course completed. Maintaining a training record will help you to track and verify that each volunteer has completed required training and any necessary refresher training. A training record can also help you know when training needs to be scheduled. Finally, in the event of a legal dispute, an up-to-date training record will show that the volunteer has received the training required to complete job tasks accurately and safely.

A sample training record is shown below.

Job Aid 3.5: Sample Training Record

Name: _____   Position: _____

Course Title	Course Code	Date Completed	Expiration Date	HR Verification

Step 6: Supervising and Evaluating Volunteers

The basic principles for "hiring" and training volunteers are the same as those that apply to paid staff. The same is true of supervision.

Good supervision empowers volunteers to succeed and ensures the completion of assigned work. A good supervisor:

- Delegates responsibilities and authority effectively.

- Establishes performance expectations.

- Acts as a coach and team builder.

- Communicates his or her ideas effectively.

- Knows how to listen, and is receptive to information from others.

- Assists staff in developing their skills.

- Gives constructive feedback and takes corrective action, when needed.

- Recognizes staff for their contributions.

The scope of this course does not permit a lengthy discussion of supervision skills. However, good resources exist for those who wish to improve their supervisory ability. (See the end of this unit for more information.)

Recognition and Volunteer Retention

Recognition is a critical component of supervision because it is one of the keys to maintaining volunteer interest and, therefore, volunteer retention. Think back to the common volunteer complaints listed in Unit 2 (p. 2.5). Most of them can be avoided by good supervision, particularly recognition. Volunteers can sometimes feel like second-class citizens when compared to paid staff. Recognition can go a long way toward alleviating a perceived hierarchy.

Recognition can be as Informal as saying "thank you," or as formal as a plaque presented at a banquet. Job Aid 3.6 on page 3.36 lists formal and informal ways to recognize and motivate volunteers. Think about which of these your agency typically does well and which it might like to try.

Recognition and Volunteer Retention (Continued)

To be effective, recognition should be:

- Sincere.

- Ongoing.

- Inclusive of *all* volunteers.

- Varied (both formal and informal).

- Continually evaluated.

- Meaningful to the individual.

- Supported by top management.

Job Aid 3.6 on the following page provides some examples of volunteer recognition.

Job Aid 3.6: Ways to Recognize and Motivate Volunteers

Informal

- Address a volunteer by name.

- Say "thank you."

- Write a thank you note.

- Say "good job."

- Treat a volunteer to coffee.

- Take him or her to lunch.

- Ask how work is going and stop to listen and discuss the response.

- Ask for input.

- Include volunteers in staff meetings.

- Include volunteers in an orientation video.

Formal

- Give annual recognition at an appreciation banquet.

- Hold an awards ceremony during National Volunteer Week.

- Throw a holiday party for volunteers.

- Place a photo and article in the local newspaper featuring volunteers.

- Place a "Volunteer of the Month" photo on the agency bulletin board.

- Present volunteers with plaques, certificates, pins, t-shirts, coffee mugs, etc.

- Ask a volunteer to serve on an advisory board.

- Offer advanced training.

- Give more responsibility, such as the opportunity to train or supervise other volunteers.

Recognition and Volunteer Retention (Continued)

While recognition is important, it is only one factor in ensuring volunteer satisfaction. Other ways to increase volunteer satisfaction include:

- Ensuring that the position and assignment are a good match for the person's abilities and interests.

- Making sure that volunteers have the resources they need to do their assigned tasks.

- Allowing volunteers to see the impact of their work, whenever possible.

- Helping to overcome logistical barriers to volunteering, such as offering on-site childcare for single parents.

- Making sure that the environment is comfortable and safe.

- Building some fun opportunities for socializing into the volunteer experience.

Evaluating Volunteers

Evaluative feedback can also be motivating for volunteers. Volunteers have the right to know what is expected of them and how well they are meeting those expectations. Positive evaluations provide the basis for volunteer recognition, while noting areas that need improvement provides an opportunity to adjust whatever is necessary to increase the volunteer's effectiveness.

There are several factors that you should consider to ensure that evaluations run smoothly and are as successful as possible.

- Establish a nonthreatening environment early in the session, and maintain it throughout.

- Begin with a discussion of the volunteer's strengths, then move to areas in which the volunteer needs improvement. Provide concrete examples of each.

- Listen to the volunteer's input and take it seriously.

Evaluating Volunteers (Continued)

- Develop a plan for improving weaknesses jointly with the volunteer. An improvement plan should include:

  - How both parties will contribute to achieving the plan's goals.
  - How the plan's success will be measured.
  - The timeframes for additional feedback.

Evaluations can be formally scheduled performance reviews. Or, if this process is seen as too threatening by the volunteer, evaluations can be done informally by observing volunteers periodically while they work and by sitting down occasionally to discuss how they feel they are doing.

Even if evaluations are done formally, there should be no surprises because the volunteer should be receiving informal feedback between performance reviews. Address problems early to avoid further difficulties later on.

If evaluations are done informally, they should still be documented, especially if there are problems with the volunteer's performance. Needless to say, all evaluations should be kept confidential.

Other guidelines for evaluations include:

- Making sure comments are fair.

- Focusing on the work, not on the individual.

- Following agency guidelines for disciplinary procedures. (Disciplinary policy should be covered during orientation.) Corrective action may include:

  - Additional training or supervision.
  - Reassignment.
  - Suspension.
  - Termination.

Termination should be reserved for those times when all other measures have failed, or when there has been gross misconduct such as theft, abuse, or being under the influence of drugs or alcohol.

During the orientation session, volunteers should have been made aware of grievance and appeal procedures that they can use if they have concerns about their evaluation or their job that cannot be resolved with their supervisor.

Evaluating Volunteers (Continued)

If a volunteer resigns, conduct an exit interview to find out:

- Why he or she is leaving.

- Suggestions for improving the position and/or volunteer experience in your agency.

Activity: Developing an Evaluation Instrument

This activity will provide an opportunity for you to develop an evaluation instrument. For the volunteer position that you selected in the previous activity, create an evaluation form in the space provided below. Develop both general criteria (such as punctuality) and criteria specific to the position.

Position: _____

General Performance Criteria:

Job-Specific Criteria:

Step 7: Evaluating Volunteer Programs

Two-way feedback during the evaluation process allows you to evaluate your volunteer program as well as your volunteers. Evaluating the volunteer program regularly ensures that it is meeting the needs of the:

- Agency.

- Community.

- Volunteers.

Program evaluation also allows you to:

- Increase volunteer satisfaction.

- Upgrade the program to improve services.

- Improve efficiency and reduce costs.

- Identify problems and solutions.

- Determine what works and what does not.

Volunteer evaluation should not be the only means of program evaluation. In addition to the volunteers themselves, other sources of feedback include:

- Paid staff.

- The public.

- Exercise performance.

- Lessons learned from actual events.

Step 7: Evaluating Volunteer Programs (Continued)

Consider whose input is important to determining the program's success, and solicit input from those groups or individuals.

As important as *who* gives feedback is *how* the feedback is gathered.

Methods for gathering and processing feedback include:

- Staff surveys.

- Internal reviews that compare program results with objectives.

- External reviews in which you compare and contrast your agency's volunteer program with other volunteer programs.

When analyzing the results of feedback, be sure to think through alternate explanations, rather than making assumptions or jumping to conclusions about whether the results are related to the program itself or the individual volunteers. Develop workable solutions to identified problems and implement them promptly to keep the program on track.

Unit Summary

This unit examined the seven steps necessary to build and maintain a strong volunteer program that meets the needs both of your agency and the individual volunteers:

- Conducting a needs analysis

- Writing volunteer job descriptions

- Recruiting

- Placing

- Training

- Supervising and evaluating

- Evaluating the program itself

The next unit will discuss coordinating with other voluntary agencies, community-based organizations, and other groups that provide volunteer services.

For More Information

- Council for Certification in Volunteer Administration (CCVA)

  http://www.cvacert.org/

- FEMA course on leadership and influence

  http://training.fema.gov/EMIWeb/IS/is240.asp

Knowledge Review

Select the best response. Then turn the page to check your answers.

1. A well-written job description is created during which of the following processes?

   a. Program Design
   b. Placement
   c. Training
   d. Supervision and evaluation

2. When you need volunteers with specific skills to do specific jobs, which type of recruitment should you use?

   a. Broad-based
   b. Targeted

3. Which of the following placement steps is a two-way process?

   a. Screening
   b. Interviewing

4. Which of the following training steps involves skills instruction?

   a. Orientation to the agency
   b. Job training

5. Asking a volunteer to serve on an advisory board is an example of which type of recognition?

   a. Informal
   b. Formal

Knowledge Review (Continued)

1. a
2. b
3. b
4. b
5. b

Unit 4: Working with VOADs, NGOs, and Other Groups

Introduction

Voluntary Organizations Active in Disaster (VOADs), nongovernmental organizations (NGOs), business and industry, and other groups have a long record of community service, both day-to-day and, especially, when an emergency threatens or strikes a community.

In this unit, you will learn about how to coordinate with VOADs, NGOs, businesses, and other groups for volunteer services. After completing this unit, you should be able to:

- List the key responsibilities of a VOAD/NGO Coordinator.

- Describe the role of VOADs in providing emergency assistance.

- Identify VOADs, NGOs, businesses, and other groups in your area that can provide emergency assistance.

- Develop a plan for working with VOADs, NGOs, businesses, and other groups.

The Role of the VOAD/NGO Coordinator

Perhaps you have determined on the basis of your community and agency needs analysis that it is impractical for your emergency management agency to develop its own volunteer program. Or perhaps you have more needs than your agency volunteers alone can cover. In either case, you will need to coordinate with volunteers affiliated with other agencies.

The role of a VOAD/NGO Coordinator, as opposed to that of a Volunteer Program Director, does not involve recruitment, training, and evaluation of individual volunteers. Instead, he or she:

- Builds relationships and acts as a liaison with local community VOADs, NGOs, and business and industry for the provision of volunteer services, when necessary.

The Role of the VOAD/NGO Coordinator (Continued)

- Collaborates with local community VOADs, NGOs, and others to develop and exercise a plan on how to coordinate volunteer services in a disaster. This plan should be formalized as a Donations/Volunteer Annex to the Emergency Operations Plan (EOP) and should provide answers to the following questions:

    - Will the community have a Volunteer Center to serve as a central clearinghouse to match requests for volunteer services with available volunteer resources? If so, who will staff the center? If not, how will volunteer services be coordinated?
    - Which agencies will be responsible for providing which emergency services (e.g., food, shelter)?
    - Who will be responsible for training and supervising the volunteers?
    - Who will be responsible for emergency public information? What should the content of the message to the public be regarding volunteering in a disaster? Which media will be used to get the message out?

Working with VOADs, NGOs, business and industry, and others to develop a plan for addressing volunteer needs *before* an emergency can help eliminate some of the problems associated with dealing with unaffiliated or spontaneous volunteers *during* an emergency, and it just makes good sense.

Coordinating With the VOAD

A good place to begin connecting with community VOADs is with the umbrella organization National Voluntary Organizations Active in Disaster (NVOAD). NVOAD is a consortium of recognized national voluntary organizations that play a critical role in disaster relief. Such organizations provide capabilities to incident management and response efforts at all levels. Its mission is to foster more effective service to people affected by disaster through cooperation, coordination, communication, education, and outreach to State VOADs. Job Aid 4.1 on page 4.5 lists the 49 member agencies of the NVOAD.

Coordinating With the VOAD (Continued)

VOAD member agencies have an active role in assisting emergency management personnel in planning for volunteer involvement during emergencies. VOAD member agencies are independent, but in planning for and responding to emergencies, the State VOAD provides the mechanism for collaboration and communications among the voluntary agencies. The State VOAD is organized during the preparedness phase, and also works through each phase of emergency management activities to ensure that the citizens' needs are met.

Working with other agencies' volunteers, as opposed to directing your own volunteers, requires interagency collaboration. Collaboration is the process in which agencies work together as a team on a common mission.

A successful collaborative relationship requires:

- A commitment to participate in *shared decisionmaking.*

- A *willingness to share* information, resources, and tasks.

- A professional sense of *respect* for individual team members.

Collaboration with other agencies and groups doesn't just "happen." Collaboration can be made difficult by differences among agencies in:

- Terminology.

- Experience.

- Priorities.

- Culture.

Collaboration under these conditions requires flexibility to agree on common terms and priorities, and the humility to learn from others' ways of doing things.

Coordinating With the VOAD (Continued)

After working through common issues and agreeing to solutions, the benefits of collaboration follow naturally. Interagency collaboration benefits the community by strengthening the overall response to emergencies. For example, collaboration:

- Eliminates duplication of services, resulting in a more efficient response.

- Expands resource availability.

- Enhances problemsolving through cross-pollination of ideas.

In other words, collaboration results in a "win-win" for both the agencies and the community.

Job Aid 4.1: NVOAD National Membership

Adventist Community Services	Mennonite Disaster Service
American Baptist Men	Mercy Medical Airlift
American Radio Relay League, Inc	National Association of Jewish Chaplains
American Red Cross	National Baptist Convention USA
Billy Graham Rapid Response Team	National Emergency Response Team
Brethren Disaster Ministries	National Organization for Victim Assistance
Catholic Charities USA	Nazarene Disaster Response
Christian Reformed World Relief Committee	Noah's Wish
Churches of Scientology Disaster Response	Operation Blessing
Church World Service	Points of Light Foundation/Hands On Network
City Team Ministries	Presbyterian Disaster Response
Convoy of Hope	REACT International, Inc.
Episcopal Relief and Development	Samaritan's Purse
Feeding America	Save The Children
Feed The Children	Society of St. Vincent DePaul
Foundation of Hope - ACTS World Relief	Southern Baptist Convention/NAMB
Habitat for Humanity International	The Salvation Army
Hands on Disaster Response	Tzu Chi Foundation
Hope Coalition America (Operation Hope)	United Church of Christ
HOPE worldwide, Ltd.	United Jewish Communities
Humane Society of the United States	United Methodist Committee On Relief
International Critical Incident Stress Foundation	United Way of America
International Relief and Development (IRD)	World Hope International
Latter-day Saint Charities	World Vision
Lutheran Disaster Response	

For more information on National Voluntary Organizations Active in Disaster, visit www.NVOAD.org.

Activity: Identifying Local VOADs

This activity will provide an opportunity for you to identify the voluntary organizations in your community. A good way to start is by contacting your State VOAD to see which VOAD members are operating in your State. Then check your local phone book to see which have an established presence in your community. Use the space below to make a list to begin the process of contact and collaboration.

Agency   **Address**   **Phone**

Coordinating with NGOs and Businesses

In addition to VOADs, nongovernmental organizations (NGOs) and businesses have played a role in disaster preparedness and response. Some, such as IBM, have already established volunteer programs. Banks are under a Federal requirement to donate a certain number of volunteer hours in community service, which makes them another potential volunteer resource for emergency management. Other groups, such as the Elks and Veterans of Foreign Wars (VFW), have a history of donating goods and services during a disaster.

Think creatively about organizations that could help to meet your needs. Other groups that are potential sources of volunteer services are professionals whose skills are known to be needed in disasters. These groups include medical professionals, such as doctors and nurses, and licensed radio operators.

Additionally, the United We Serve initiative is a potential source of volunteers. (See the Web site listed under "For More Information" at the end of this unit.)

Under this initiative, a national network of State, local, and Tribal Citizen Corps Councils brings together leaders from law enforcement, fire, emergency medical, and other emergency management volunteer organizations, local elected officials, the private sector, and other community stakeholders.

Local Citizen Corps Councils implement Citizen Corps programs, including Community Emergency Response Teams (CERTs), which are potential sources of emergency assistance.

CERTs are neighborhood and workplace volunteers who train together to develop emergency response skills. They apply these skills to help others following a major disaster when professional help is not readily available or stretched thin. CERT members work with emergency management and become part of the emergency response capability for the area in which they live.

Areas covered during training include individual and family emergency preparedness, emergency medical response, fire safety, light search and rescue, team organization and disaster psychology. CERT members practice their skills by taking part in hands-on exercises and activities.

Communities using CERTs have important advantages in recruiting, training, and maintaining emergency management volunteers. To learn more about CERT volunteers and also to see examples of the other valuable community services they provide, go to http://citizencorps.gov/cert.

Coordinating with NGOs and Businesses (Continued)

Local Councils implement other programs as well, including:

- Medical Reserve Corps

- Fire Corps

- USAonWatch – Neighborhood Watch

- Volunteers in Police Service

- Affiliate Programs.

These programs provide opportunities for special skills and interests, develop targeted outreach for special-needs groups, and organize special projects and community events.

Citizen Corps Affiliate Programs expand the resources and materials available to States and local communities through partnerships with programs and organizations that:

- Offer resources for public education, outreach, and training.

- Represent volunteers interested in helping to make their communities safer.

- Offer volunteer service opportunities to support first responders, disaster relief activities, and community safety efforts.

Other programs unaffiliated with Citizen Corps also provide organized citizen involvement opportunities in support of Federal response to major disasters and events of national significance. One example is the National Animal Health Emergency Response Corps (NAHERC), which helps protect public health by providing a ready reserve of private and State animal heath technicians and veterinarians to combat threats to U.S. livestock and poultry in the event of a large outbreak of a foreign animal disease.

Coordinating with NGOs and Businesses (Continued)

Think creatively about how emergency needs can be met through VOADs, NGOs, and business and industry. Virtually all of the groups are more than willing to help—if they only knew what to do.

Other organizations, such as the Points of Light Foundation, have valuable resources for emergency managers looking to involve corporate volunteers:

- **Corporate members** are companies that have employee volunteer programs, which have been initiated or are maintained through the help of the Points of Light Foundation.

- **Corporate Volunteer Councils** (CVCs) are coalitions of local businesses and corporations that either have active employee and/or retiree volunteer involvement programs, or are interested in initiating such programs. In different parts of the country, CVCs may also be known as Business Volunteer Councils (BVCs), or by other names.

- **The National Council on Workplace Volunteerism** promotes volunteerism in and through the business community on behalf of the Points of Light Foundation by leading and supporting the development and growth of employee volunteer programs and Corporate Volunteer Councils.

The best point of contact for finding corporate volunteers is usually the company's public relations director.

Unlike VOAD volunteers, corporate volunteers may not be trained in disaster response. Therefore, coordination with these volunteers may involve training them in advance. FEMA, the Points of Light Foundation, and The American Red Cross offer training resources to assist in this task.

Case Study: Ice Storm

Sometimes, it may be necessary to draw on all available resources to meet emergency needs. Review the case study below and develop some ideas for addressing volunteer needs.

A severe ice storm has struck a large Midwestern city. Several inches of ice have formed on tree limbs and power lines. Tree branches have fallen by the thousands, causing widespread power outage to tens of thousands of people for several days. Power companies are doing only minimal service to restore power to homes—no debris removal or tree trimming. Tree trimming services are quoting waiting periods of a month or more for emergency tree service. Chainsaw teams and individuals to clear tree limb debris are desperately needed.

City and county managers have established more than 60 shelters—the largest shelter operation in decades. The American Red Cross and The Salvation Army are overwhelmed with the mass feeding and sheltering needs and can only support a few of the total numbers of shelters. Staff shortages are the biggest problem. Their core volunteer staff is very well trained, but they are rapidly becoming worn out.

In addition to mass care needs, special needs populations are seriously threatened. Nursing homes, in-home health care patients, people with disabilities, and seniors are in distress. They are calling 9-1-1, churches, and neighbors for help. These people are, in turn, looking to local government for help.

The regional and national news media are carrying little or no information about the ice storm. The local media have an intense focus on the storm, but many people are unable to view their television as a result of power outages.

Very few people are calling to offer help. The United Way Volunteer Coordination Center has reported only a handful of calls, probably because of the power outage in so many homes and the resulting lack of widespread media viewing.

The city Emergency Operations Center (EOC) is activated, but so far only city government agencies have had a consistent presence.

What can city emergency management do to address the burgeoning need for volunteer resources in the community?

Case Study: Ice Storm (Continued)

Answers to the Case Study

What can city emergency management do to address the burgeoning need for volunteer resources in the community?

Some possible answers include:

- Convene key voluntary and human services agencies in the EOC to begin identifying needs and resources.
- Request help from the local, State, or national VOAD to help meet the needs.
- Use the local Volunteer Center to recruit local volunteers from NGOs and corporate volunteer programs.
- Use the media, especially print media and radio, to publicize the need for volunteers.

Note: The city should not organize volunteers to remove debris unless it has its own volunteer program and can assume the legal liability for such work. (Legal issues are covered in Unit 5.)

Activity: Identifying Local NGOs and Businesses

This activity will provide an opportunity for you to identify the NGOs and businesses in your community that have provided or could provide volunteer services in an emergency. Begin with the Points of Light Foundation's corporate members (a list is available on their Web site), then check your local phone book to see which of these companies have offices in your community. Continue by checking with local community organizations (such as the Elks or Veterans of Foreign Wars) and professional associations for established volunteer programs, or those with a history or interest in providing volunteer services in a disaster. Then use the space below to make a list of those organizations that may be resources in a time of disaster and identify the services they may be able to provide.

Organization   **Point of Contact**   **Phone**

Activity: Developing a Strategy for Working With VOADs, NGOs, Businesses, and Other Groups

*This activity will provide an opportunity for you to develop a strategy to collaborate with community organizations for volunteer services. Contact the VOADs on the list you developed in a previous activity and invite representatives to a meeting to develop (or update) the Donations/Volunteer Annex to your EOP. Use the space below to guide the discussion at the meeting. Be sure to address the topics listed on page 4.1 (e.g., emergency public information), as well as how to coordinate with potentially untrained NGO and corporate volunteers. (**Note:** Unaffiliated volunteers, covered in the next unit, should also be addressed.)*

Primary Agency:

Support Agencies:

Activity: Developing a Strategy for Working With VOADs, NGOs, Businesses, and Other Groups (Continued)

Introduction:

Purpose:

Scope:

Policies:

Situation and Assumptions:

Activity: Developing a Strategy for Working With VOADs, NGOs, Businesses, and Other Groups (Continued)

Concept of Operations:

Roles and Responsibilities:

Activity: Developing a Strategy for Working With VOADs, NGOs, Businesses, and Other Groups (Continued)

Resource Requirements:

Essential Emergency Information:

Activity: Developing a Strategy for Working With VOADs, NGOs, Businesses, and Other Groups (Continued)

References:

Terms and Definitions:

Mandates: Emergency Management and Coordination Systems

On February 28, 2003, the President issued Homeland Security Presidential Directive 5 (HSPD–5), "Management of Domestic Incidents," which directed the Secretary of Homeland Security to develop and administer a National Incident Management System (NIMS). This system provides a consistent nationwide template to enable Federal, State, tribal, and local governments, nongovernmental organizations (NGOs), and the private sector to work together to prevent, protect against, respond to, recover from, and mitigate the effects of incidents, regardless of cause, size, location, or complexity. This consistency provides the foundation for utilization of NIMS for all incidents, ranging from daily occurrences to incidents requiring a coordinated Federal response.

National Incident Management System (NIMS)

NIMS is not an operational incident management or resource allocation plan. NIMS represents a core set of doctrines, concepts, principles, terminology, and organizational processes that enables effective, efficient, and collaborative incident management.

Building on the foundation provided by existing emergency management and incident response systems used by jurisdictions, organizations, and functional disciplines at all levels, NIMS integrates best practices into a comprehensive framework for use nationwide by emergency management/response personnel in an all-hazards context. These best practices lay the groundwork for the components of NIMS and provide the mechanisms for the further development and refinement of supporting national standards, guidelines, protocols, systems, and technologies. NIMS fosters the development of specialized technologies that facilitate emergency management and incident response activities, and allows for the adoption of new approaches that will enable continuous refinement of the system over time.

NIMS (Continued)

According to the National Integration Center, "institutionalizing the use of ICS" means that government officials, incident managers, and emergency response organizations at all jurisdictional levels adopt the Incident Command System. Actions to institutionalizing ICS takes place at two levels—policy and organizational/operational:

At the policy level, institutionalizing ICS means government officials:

- Adopt ICS through executive order, proclamation or legislation as the jurisdiction's official incident response system.

- Direct that incident managers and response organizations in their jurisdictions train, exercise, and use ICS in their response operations.

At the organizational/operational level, incident managers and emergency response organizations should:

- Integrate ICS into functional, system-wide emergency operations policies, plans, and procedures.

- Provide ICS training for responders, supervisors, and command-level officers.

- Conduct exercises for responders at all levels, including responders from all disciplines and jurisdictions.

NIMS integrates existing best practices into a consistent, nationwide approach to domestic incident management that is applicable at all jurisdictional levels and across functional disciplines. Five major components make up the NIMS system approach:

NIMS (Continued)

Five major components make up the NIMS system approach:

- **Preparedness:** Effective emergency management and incident response activities begin with a host of preparedness activities conducted on an ongoing basis, in advance of any potential incident. Preparedness involves an integrated combination of assessment; planning; procedures and protocols; training and exercises; personnel qualifications, licensure, and certification; equipment certification; and evaluation and revision.

- **Communications and Information Management:** Emergency management and incident response activities rely on communications and information systems that provide a common operating picture to all command and coordination sites. NIMS describes the requirements necessary for a standardized framework for communications and emphasizes the need for a common operating picture. This component is based on the concepts of interoperability, reliability, scalability, and portability, as well as the resiliency and redundancy of communications and information systems.

- **Resource Management:** Resources (such as personnel, equipment, or supplies) are needed to support critical incident objectives. The flow of resources must be fluid and adaptable to the requirements of the incident. NIMS defines standardized mechanisms and establishes the resource management process to identify requirements, order and acquire, mobilize, track and report, recover and demobilize, reimburse, and inventory resources.

- **Command and Management:** The Command and Management component of NIMS is designed to enable effective and efficient incident management and coordination by providing a flexible, standardized incident management structure. The structure is based on three key organizational constructs: the Incident Command System, Multiagency Coordination Systems, and Public Information.

- **Ongoing Management and Maintenance:** Within the auspices of Ongoing Management and Maintenance, there are two components: the National Integration Center (NIC) and Supporting Technologies.

Additional information about NIMS can be accessed online at http://www.fema.gov/emergency/nims/ or by completing EMI's IS 700 online course.

National Response Framework (NRF)

The NRF is a guide to how the Nation conducts all-hazards response – from the smallest incident to the largest catastrophe. This key document establishes a comprehensive, national, all-hazards approach to domestic incident response. The Framework identifies the key response principles, roles, and structures that organize national response. It describes how communities, States, the Federal Government, and private-sector and nongovernmental partners apply these principles for a coordinated, effective national response.

The NRF is:

- **Always in effect, and elements can be implemented as needed on a flexible, scalable basis to improve response.** It is not always obvious at the outset whether a seemingly minor event might be the initial phase of a larger, rapidly growing threat. The NRF allows for the rapid acceleration of response efforts without the need for a formal trigger mechanism.

- **Part of a broader strategy.** The NRF is required by, and integrates under, a larger National Strategy for Homeland Security that:

    - Serves to guide, organize, and unify our Nation's homeland security efforts.

    - Reflects our increased understanding of the threats confronting the United States.

    - Incorporates lessons learned from exercises and real-world catastrophes.

    - Articulates how we should ensure our long-term success by strengthening the homeland security foundation we have built.

- **Comprised of more than the core document.** The NRF is comprised of the core document, the Emergency Support Function (ESF), Support, and Incident Annexes, and the Partner Guides. The core document describes the doctrine that guides our national response, roles and responsibilities, response actions, response organizations, and planning requirements to achieve an effective national response to any incident that occurs.

    The following documents provide more detailed information to assist practitioners in implementing the Framework:

    - **Emergency Support Function Annexes** group Federal resources and capabilities into functional areas that are most frequently needed in a national response (e.g., Transportation, Firefighting, Search and Rescue).

NRF (Continued)

- **Support Annexes** describe essential supporting aspects that are common to all incidents (e.g., Financial Management, Volunteer and Donations Management, Private-Sector Coordination).

- **Incident Annexes** address the unique aspects of how we respond to seven broad incident categories (e.g., Biological, Nuclear/Radiological, Cyber, Mass Evacuation).

Additional information about the NRF can be accessed online at http://www.fema.gov/emergency/nrf or by completing EMI's IS 800.b online course.

Volunteer and Donations Management Support Annex

The Volunteer and Donations Management Support Annex of the NRF provides guidance on the Federal role in supporting State governments in the management of masses of unaffiliated volunteers and unsolicited donated goods.

This annex describes the coordination processes used to support the State in ensuring the most efficient and effective use of unaffiliated volunteers, unaffiliated organizations, and unsolicited donated goods to support all Emergency Support Functions (ESFs) for incidents requiring a Federal response, including offers of unaffiliated volunteer services and unsolicited donations to the Federal Government.

This guidance applies to all agencies and organizations with direct and indirect volunteer and/or donations responsibilities under the National Response Framework.

The Volunteer and Donations Management Support Annex can be accessed at http://www.fema.gov/pdf/emergency/nrf/nrf-support-vol.pdf

What This Means to You

Your jurisdiction is required to:

- Use NIMS to manage all incidents, including recurring and/or planned special events.

- Integrate all response agencies and entities into a single, seamless system, from the Incident Command Post, through department Emergency Operations Centers (DEOCs) and local Emergency Operations Centers (EOCs), through the State EOC to the regional- and national-level entities.

- Develop and implement a public information system.

- Identify and type all resources according to established standards.

- Ensure that all personnel are trained properly for the job(s) they perform.

- Ensure the interoperability and redundancy of communications.

Remember the importance of working with VOADs, NGOs, business and industry, and others to develop a plan for addressing volunteer needs *before* an emergency to help eliminate some of the potential problems that can occur *during* an emergency.

Unit Summary

This unit examined how to identify and collaborate with local voluntary agencies, community-based organizations, businesses, and other groups to obtain volunteer services in a disaster. The next unit will discuss special issues involved in working with volunteers.

For More Information

- Citizen Corps

  http://www.citizencorps.gov/

- Points of Light Foundation

  http://www.pointsoflight.org/

  Offers a complete list of Volunteer Centers and State Offices of Volunteerism.

- United We Serve

  http://www.serve.gov/

  This government-sponsored Web site gives information on President Obama's United We Serve initiative.

- Your local Volunteer Center

  1-800-VOLUNTEER

- National Incident Management System (NIMS)

  http://www.fema.gov/emergency/nims/

- National Response Framework (NRF)

  http://www.fema.gov/emergency/nrf/

- Volunteer and Donations Management Support Annex

  http://www.fema.gov/pdf/emergency/nrf/nrf-support-vol.pdf

 **Knowledge Review**

Select or provide the best answer. Turn the page to check your answers against the solutions.

1. A VOAD/NGO Coordinator recruits, trains, and supervises VOAD volunteers.

   a. True
   b. False

2. NVOAD members include both VOADs and NGOs.

   a. True
   b. False

3. Collaboration involves sharing information and resources.

   a. True
   b. False

4. Corporate volunteers usually have been trained in disaster response.

   a. True
   b. False

5. The best point of contact for corporate volunteers is the company's vice president.

   a. True
   b. False

Knowledge Review (Continued)

1. b
2. b
3. a
4. b
5. b

Unit 5: Special Issues

Introduction

In this unit, you will learn about special issues involved in running volunteer programs. After completing this unit, you should be able to:

- Develop a plan for dealing with spontaneous volunteers.

- Identify special issues involved with managing volunteers and local points of contact for advice in each area.

- Develop a plan for managing and reducing volunteer stress.

Note: It is not possible to identify all of the special issues that may arise in managing volunteers. This unit will point to only the issues that communities have identified as posing the greatest challenges. While the unit will suggest solutions, they will necessarily be generic, because specific solutions must take into account State and local laws, ordinances, and policies.

Dealing With Unaffiliated Volunteers

Unaffiliated volunteers, also known as spontaneous volunteers, are individuals who offer to help or self-deploy to assist in emergency situations without fully coordinating their activities.
Let's begin examining this issue with three scenarios that illustrate the potential benefits and challenges of spontaneous volunteers.

Scenario 1

A boat carrying 50 passengers and crew on a large lake in the Midwest is caught in a heavy storm and capsizes. Approximately a dozen nearby boaters rush to the overturned vessel. Some dive repeatedly under the boat and help locate and free many people who are trapped underneath. Others pull people from the water and bring them to shore. All of the victims are rescued before professional first responders even reach the site.

Dealing With Unaffiliated Volunteers (Continued)

Scenario 2

In response to a flash flood disaster, volunteers are sent out to conduct a needs assessment in the community. Two of the volunteers meet a family that apparently needs food, clothing, and other essentials. Touched by the family's plight, the workers "adopt" the family for the afternoon. The volunteers take the family to a local shopping center where they pay for food and clothing out of their own pockets. As it turned out, the family had no damages from the flood but was typical of hundreds of families from this economically depressed area of the State.

Scenario 3

A plane hits a tall building filled with office workers. People are escaping the burning building into the streets. A chiropractor's assistant is walking down the street and sees the people running toward him. He hurries toward those in need and helps a woman who had fallen. He continues to help people flee until they are out of danger. Then he joins a construction crew that pulls debris from the street so the emergency vehicles can get through. For the next few days, he does not go to his paid job, but continues to work tirelessly at various volunteer jobs to help the victims of this disaster. This man goes on to become a paid staff member for a voluntary organization.

As illustrated by the first scenario, unaffiliated volunteers present potential benefits:

- They supply extra (or immediate) assistance that may be needed in an emergency. For example, sometimes lots of people are needed for unskilled work, such as sandbagging rising streams. Thus they can fill in when the response is shorthanded.

- They represent a willing workforce, some of whom may go on to become trained, affiliated volunteers who can be counted on in future disasters.

Unaffiliated volunteers have not been screened (i.e., had background checks). Thus, some "volunteers" may have motives other than helping. They may be potential terrorists, or at least "rip-off artists." Sometimes these "volunteers" show up repeatedly following successive disasters. Those volunteers who are honest (and it's not always easy to tell the difference until it is too late) may need close supervision because of their lack of training—which was the challenge illustrated by the second scenario.

Dealing With Unaffiliated Volunteers (Continued)

However, unaffiliated volunteers also present significant challenges. Unaffiliated volunteers are not associated with or trained by a voluntary organization. (Volunteers who are registered with a voluntary agency are instructed *not* to self-deploy.) Unaffiliated spontaneous volunteers pose potentially serious issues for the response, which has become especially clear since September 11, when thousands showed up to "help" in a very dangerous situation.

The issue of unaffiliated volunteers rose to a new level of importance after the terrorist attacks of September 11. People—professional and unskilled, from in-State, out-of-State, and even Canada—showed up in unprecedented numbers. Although these people were well intentioned, their numbers created a logistical nightmare for professional emergency responders and volunteer coordinators. Not only were there not enough jobs for all of the untrained volunteers, but they:

- Created a physical logjam that interfered with the response.

- Were not cognizant of acting carefully so as to preserve evidence at the crime scene.

- Created extra work for some responders who had to deal with directing (or rescuing) volunteers instead of being able to focus on their primary job of incident response.

- Had the potential for creating additional danger and casualties—to themselves and others—because of their lack of training (e.g., in search and rescue).

Even though spontaneous volunteers can potentially create problems if they "just show up," it is best to have a plan for how to deal with them—because they will show up. Any major disaster will bring out spontaneous volunteers who are unaffiliated. And in all fairness, they can be valuable assets if you have a system for using them.

In one documented case in Florida, the community used unaffiliated volunteers for cleanup after a tornado. Osceola County completed its cleanup 35 days sooner, and spent $6.6 million less than initial estimates because of volunteer help. As a bonus, by keeping records of volunteer hours for categories of work, the county could apply volunteer work as the major portion of its matching share requirement for FEMA disaster assistance.

Dealing With Unaffiliated Volunteers (Continued)

Some good ideas are available in the pamphlet, *Unaffiliated Volunteers in Response and Recovery* by the Florida Commission on Community Service. Key among these is to appoint a volunteer coordinator to be included in the predisaster planning process and to designate a location that can serve as a volunteer reception center. Steps in Unit 3 for dealing with individual volunteers are generally useful for spontaneous volunteers after a disaster, but in a compressed timeframe.

This Volunteer and Donations Management Support Annex provides guidance on the Federal role in supporting State governments in the management of masses of unaffiliated volunteers and unsolicited donated goods. This guidance applies to all agencies and organizations with direct and indirect volunteer and/or donations responsibilities under the National Response Framework.

The Federal Government encourages State, tribal, and local governments to coordinate with voluntary agencies, community and faith-based organizations, volunteer centers, and private-sector entities through local Citizen Corps Councils and local Voluntary Organizations Active in Disaster (VOADs) to participate in preparedness activities including planning, establishing appropriate roles and responsibilities, training, and exercising.

Private nonprofit and private-sector organizations that can provide a specific disaster-related service to Federal, State, local, and tribal governments are encouraged to establish preincident operational agreements with emergency management agencies. At the Federal level, FEMA will provide preincident support to broker a match with the most appropriate ESF or response element for organizations with disaster services that are not currently affiliated with a specific ESF.

State, tribal, and local governments have primary responsibility, in coordination with VOADs, to develop and implement plans to manage volunteer services and donated goods. DHS/FEMA recommends that States and local jurisdictions develop and strengthen a Volunteer and Donations Management ESF/Support Annex in their State and local emergency plans.

These plans should detail volunteer and donations management-related outreach and education programs, procedures to activate mutual aid such as the Emergency Management Assistance Compact, communications and facilities management, a Volunteer/Donations Coordination Team, a call center, relevant points of contact, safety and security, and demobilization.

Planning should be a cooperative effort among all of the VOADs and emergency management/response agencies in your community and should include the following elements:

Dealing With Unaffiliated Volunteers (Continued)

- The role of *emergency public information* cannot be overemphasized in dealing with unaffiliated volunteers. The media can and should be used in an emergency to discourage unaffiliated volunteers from simply showing up at the disaster site or the Emergency Operations Center (EOC). The media can publish or broadcast a number for potential volunteers to call to offer their services. A 2-1-1 phone system, such as was used for the 2002 Olympics, could be set up for potential volunteers to access volunteer organization information and referral lines. Or, if your community has a Volunteer Center, it could serve as a central clearinghouse for matching volunteer offers and requests. (If your community does not have a Volunteer Center, the United Way can fill this function.) The important point is to remove the burden of volunteer coordination from emergency managers and responders in a crisis when they have urgent, critical tasks to perform to save lives.

- For those unaffiliated volunteers who don't get the public message, an on-site *volunteer check-in area,* overseen by a Volunteer Coordinator (who should have ID and/or clothing that identifies him or her as such), can be an effective way to channel volunteers. The volunteer check-in area (identified by a large sign) should be supplied with a user-friendly skills survey for volunteers to fill out when they check in. Job Aid 5.1 on page 5.7 presents a sample skills survey. Depending on the volunteer's skills and the need, spontaneous volunteers can then be screened by the Volunteer Coordinator and given color-coded passes depending on whether they are:

  - Assigned a skilled task.
  - Given instructions and assigned a low-skill task.

Dealing With Unaffiliated Volunteers (Continued)

Note: Color-coding (or formatting) can also be used for granting access to specific areas. Be sure to work out questions about passes *before* an emergency occurs to eliminate confusion and wasted time during the response.

Volunteers who are not needed should be tactfully sent home to avoid their potential interference with the response effort. However, if possible, their information (i.e., name, address, and phone number) should be saved and entered into a database. Someone can then follow up after the emergency to recruit these individuals to become affiliated and trained volunteers.

Another option in dealing with unaffiliated volunteers is to declare a moratorium on them. Before taking this option, those involved in coordinating and assigning volunteers should weigh the pros and cons carefully (e.g., improved security vs. the loss of perhaps badly needed supplementary help). People who call in can be encouraged to give cash to a recognized VOAD instead of their time.

Sometimes groups of people will volunteer (e.g., church groups or professional clubs). The procedures for handling individual volunteers should also apply to groups.

Job Aid 5.1: Skills Survey

Name _____Address _____

Please indicate the areas that apply to you and return this survey to the Volunteer Coordinator.

PLEASE CHECK ANY OF THE FOLLOWING IN WHICH YOU HAVE EXPERTISE & TRAINING. CIRCLE YES OR NO, WHERE APPROPRIATE.

_____ First Aid (current card yes/no)   _____ CPR (current   _____ Triage   _____ Firefighting
                                                                            yes/no)

_____ Construction (electrical, plumbing, carpentry, etc.)   _____Running/Jogging

_____ Emergency Planning   _____ Emergency Management   _____ Search & Rescue

_____ Law Enforcement   _____ Bi/Multi-lingual (what language(s)) _____

_____ Mechanical Ability   _____ Structural Engineering   _____ Bus/Truck Driver
                                                                            (Commercial Driver's License)

_____ Shelter Management   _____ Survival Training &   _____ Food Preparation
                                  Techniques

_____ Ham Radio Operator   _____ CB Radio   _____ Journalism

_____ Camping   _____ Waste Disposal   _____ Recreational Leader

DO YOU HAVE EQUIPMENT OR ACCESS TO EQUIPMENT OR MATERIALS THAT COULD BE USED IN AN EMERGENCY? _____ YES _____ NO
PLEASE LIST EQUIPMENT AND MATERIALS.

_____

_____

COMMENTS

_____

_____

Addressing Legal Issues

Note: Information provided in this section is not intended to replace legal advice from counsel. It is meant to help you become aware of and understand how the law might affect volunteers in emergency management. Proper management practices can help you to minimize legal problems. *You should consult your community's legal advisor whenever you have questions.*

Safety, Risk Management, and Liability

The law generally requires that you act with the level of care that a reasonable person would exercise to prevent harm to:

- The people you are trying to help.

- Your staff—both volunteer and paid.

Volunteers for government agencies may be subject to the Federal tort claims act. A *tort* is an act that harms another person, whether intentional (e.g., assault) or unintentional (e.g., negligence). Most lawsuits are based on injuries resulting from negligence. Negligence is defined as not acting with the care that a reasonable person would have used.

If a volunteer assisting in the disaster response accidentally harms a person or their property, the law generally recognizes that the employer is responsible for the actions and inactions of his or her employee. The term "employee" covers both paid and unpaid staff. Thus, losses that result from employee actions are the responsibility of the employer.

The Volunteer Protection Act of 1997 provides legal immunity for volunteers working in disaster-related functions who are working within the scope of their assigned responsibilities, are acting in good faith, and are not guilty of gross negligence. (See the For More Information section at the end of this unit for a web address that provides information on the Volunteer Protection Act.)

Safety, Risk Management, and Liability (Continued)

Despite the Volunteers Protection Act, agencies and/or communities can be liable for volunteer actions if the agency or community has not taken steps to minimize its risk. The keys to involving volunteers successfully and minimizing susceptibility to lawsuits involve:

- **Place.**

  - First, screen all volunteers and register them as disaster service workers to cover them under Workers' Compensation and protect them from liability under the relevant provisions of the Volunteer Protection Act.
  - Follow legislation that mandates the amount of training that must be given to perform certain tasks (e.g., CPR or fork-lift operation) and the frequency with which refresher training must be given.
  - Match volunteers to tasks according to their skills and interests.
  - Provide clearly defined job descriptions and Standard Operating Procedures (SOPs).

- **Supervising.**

  - Every volunteer should have a designated supervisor.
  - Volunteers should know the limits of their authority (i.e., what they can and cannot do without specific authorization). These limits should be written in the job description and stated clearly on-site.
  - Volunteers should know the locally applicable standards of performance that they must follow.
  - Supervisors should ensure that volunteers have the proper equipment and resources to do their jobs.

- **Documenting.** Establish volunteer records, including:

  - Training received, including number of hours, results of tests, and refresher course dates.
  - Exercise participation.
  - Evidence of any necessary certification.
  - Duties and times when the volunteer was officially working.

Records should be maintained and updated regularly.

Insurance and Workers' Compensation

Despite legal protection from liability, in certain circumstances, insurance is still needed. There are different types and levels of liability insurance:

- Insurance for nonprofit organizations includes:

  - General liability.
  - Automobile.
  - Directors/officers.
  - Personal injury.

- Insurance for individuals includes:

  - Personal automobile.
  - Malpractice policies.
  - Personal injury.

In addition, some States provide Workers' Compensation coverage to *registered* disaster services workers, placing them under the umbrella of protection from on-the-job injury that employees enjoy. To be protected, all volunteers must be signed up with an established agency. Again, check your State's laws.

Activity: Consulting With Legal Counsel

Note: This activity should be completed by Volunteer Program Directors only.

Identify and meet with a local point of contact to get advice on how to handle legal issues related to volunteer issues. A place to start in researching State laws may be with your State Office of Volunteerism (if your State has one). Be sure to ask about:

- *Documenting hiring, training, evaluation, and termination procedures.*
- *Liability.*
- *Insurance and Workers' Compensation.*

Managing Volunteer Stress

During the course of performing their assigned duties, some volunteers may witness scenes that can cause extreme stress reactions, including:

- Death and injury.

- Property devastation.

- Extreme emotional reactions of victims.

In addition, disaster response work often takes place under less than ideal working conditions. Long hours and skipped meals can contribute to volunteer stress.

However, there are steps you can take before, during, and after an emergency to manage stress:

- **Before:** Ask seasoned emergency responders to talk about how they have dealt with stress during volunteer orientation and/or in special stress management training seminars.

- **During:** Supervisors should ensure that volunteers are appropriately matched to their job assignments, get regular meals and breaks, and are rotated out at the end of a reasonable-length shift.

- **After:** Invite a mental health professional to hold a Critical Incident Stress Debriefing (CISD). Critical Incident Stress Debriefings involve gathering together people who were involved in a crisis to discuss their reactions with their peers. Critical Incident Stress Management (CISM) is designed specifically for emergency response workers to:

  - Mitigate the impact of a critical incident on personnel.
  - Accelerate recovery in people who are experiencing normal reactions to abnormal events.

All of these types of activities comprise what is called Critical Incident Stress Management (CISM). (See the For More Information section of this unit for web addresses to help you understand CISM and CISD better.)

Remember that stress, which is already a part of everyone's life, is usually compounded in a disaster situation. Give your volunteers the resources they need to handle it.

Unit Summary

This unit examined special issues in volunteer management including spontaneous volunteers, legal issues, and stress management.

The next unit will summarize the key points from the course.

For More Information

- Your State office of volunteerism and/or legal counsel.

- International Critical Incident Stress Foundation, Inc.

  http://www.icisf.org/

- Understanding the Volunteer Protection Act and a summary of State volunteer protection laws

  http://www.nonprofitrisk.org

Knowledge Review

Select or provide the best answer. Turn the page to check your answers.

1. The best way to deal with spontaneous volunteers is:

   a. Before a disaster happens.
   b. At the disaster site.

2. Unaffiliated volunteers are also sometimes called _____ volunteers.

3. The law holds employers responsible for their employees' (including volunteers) actions.

   a. True
   b. False

4. The "Good Samaritan" law that limits volunteers' liability is a Federal law.

   a. True
   b. False

5. Stress is a normal reaction in a disaster situation.

   a. True
   b. False

Knowledge Review (Continued)

1. a
2. *Spontaneous*
3. a
4. b
5. a

Unit 6: Course Summary

Introduction

This unit will review the course content. After completing this unit, you should be able to summarize the key points of this course.

Volunteers and Emergency Management

A *volunteer* is any individual accepted to perform services by the lead agency (which has authority to accept volunteer services) when the individual performs services without promise, expectation, or receipt of compensation for services performed. See 16 U.S.C. 742f(c) and 29 CFR 553.101.

Volunteers can be:

- Affiliated with a voluntary organization.

- Unaffiliated, also known as spontaneous, volunteers.

A *Voluntary Organization Active in Disaster* (or *VOAD*) is an established organization whose mission is to provide services to the community through the use of trained volunteers.

There are benefits to involving volunteers. Volunteers can:

- Provide services more cost effectively.

- Provide access to a broader range of expertise and experience.

- Increase paid staff members' effectiveness by enabling them to focus their efforts where they are most needed or by providing additional services.

- Provide resources for accomplishing maintenance tasks, or upgrading that which would otherwise be put on the back burner while immediate needs demand attention.

- Enable the agency to launch programs in areas in which paid staff lack expertise.

Volunteers and Emergency Management (Continued)

- Act as liaisons with the community to gain support for programs.

- Provide a direct line to private resources in the community.

- Facilitate networking.

- Increase public awareness and program visibility.

Challenges to working with volunteers often involve misperceptions or factors, such as tension between volunteers and paid staff, that can be alleviated with good planning and management. There are, however, real obstacles to involving volunteers, such as a sparse population in rural areas.

Developing a Volunteer Program

Developing or maintaining an agency volunteer program requires someone to act as Volunteer Program Director. Developing a solid volunteer program involves seven steps:

1. Agency and Program Needs Analysis

2. Writing Volunteer Job Descriptions

3. Recruiting Volunteers

4. Placing Volunteers

5. Training Volunteers

6. Supervising and Evaluating Volunteers

7. Program Evaluation

Step 1: Agency and Program Needs Analysis

Before deciding to develop your own agency volunteer program, consider the needs of:

- Potential volunteers. Volunteer needs include the desire to make a difference, work with other people, and learn new skills.

- Your agency.

Step 1: Agency and Program Needs Analysis (Continued)

Analyzing agency needs involves:

- Considering the agency's mission.

- Looking at current staffing resources and areas of shortfall where volunteers could help.

- Describing the jobs that volunteers could do, and the abilities and resources needed to do those jobs.

Writing a well-defined job description is an important step because a good job description is:

- The first step in the recruitment process.

- A tool for marketing the agency's need to potential volunteers.

- A focal point for the interview and screening process.

- The basis for performance evaluation.

Remember to involve paid staff when planning for volunteer involvement. Soliciting their input on volunteer job descriptions will help alleviate tensions between paid and unpaid staff later on.

Step 2: Writing Volunteer Job Descriptions

Once you have completed your agency needs analysis, the next step is to write volunteer job descriptions.

When developing a job description, think about:

- The purpose of the job.

- The job responsibilities.

- Job qualifications.

- To whom the volunteer will report.

- Time commitment required for the job.

- The length of the appointment.

- Who will provide support for the position.

- Development opportunities.

Step 3: Recruitment

Recruitment can be broad-based (general) or targeted (selective). Use broad-based recruitment when there is a need for a large number of individuals for jobs that require commonly possessed skills. Use targeted recruitment when you are looking for volunteers with specific skills to do specific jobs.

Investigate the marketplace of potential volunteers in your community. Then:

- Design a message based on what you know about the volunteers you are trying to reach.

- Choose a medium (e.g., radio, television, Internet, direct mail) that will reach your targeted audience.

- Determine where and when to deliver your message.

An effective recruitment message contains the following elements:

- An interesting opening.

- A statement of the need or problem.

- A statement of how this job can meet the need or solve the problem.

- A statement to address the listener's question as to whether he or she can potentially do this job.

- How doing this job will benefit the volunteer.

- A contact point for more information.

Step 4: Placement

Placing volunteers includes:

- Screening.

- Interviewing.

All volunteers should be screened. Those jobs that involve a higher risk, such as working with money or children, require more intense screening.

Common screening tools include the:

- Application.

- Reference check.

- Professional license, criminal background, and child abuse clearance checks.

In addition, the interview is also a screening tool. If possible, the interviewer should be a volunteer, and the following tools should be used when conducting an interview:

- The potential volunteer's application

- A form for recording the interview

- A list of open-ended questions

- Information about current volunteer opportunities

The goals of the interview should be to:

- Determine the applicant's skills and motivation for volunteering.

- Match the applicant with a position.

- Answer the applicant's questions.

- Identify undesirable candidates.

Step 4: Placement (Continued)

Do not ask personal questions during an interview. Legally, you may not ask anything that is not directly related to the ability of the applicant to perform the specific volunteer job.

The interview should result in a recommendation for further action—either for placement, further screening, or rejection.

Step 5: Training Volunteers

Good training is critical because it can save lives and protect your agency from lawsuits.

There are two levels of training:

- To the agency

- To the specific job

Orientation to the agency should cover:

- The agency's mission.

- The agency's values.

- Agency procedures and issues.

- The role of volunteers in the agency.

These components can be addressed through a group presentation of a videotape and/or by various speakers.

Training for the specific job should cover:

- Specific job responsibilities.

- Who the immediate supervisor is and his or her expectations.

- Authority and accountability (i.e., what the volunteer can and cannot do without explicit direction).

- Other team members' roles.

- Resources available to do the job.

Step 5: Training Volunteers (Continued)

Training for the job is best accomplished by one-on-one mentoring.

Unlike orientation, which is information-based, training is skills-based. Like orientation, training can be general or specific. General training includes training in such skills as communication, team building, problemsolving and decisionmaking, leadership and supervision, and stress management. Specific training teaches job-specific skills, such as CPR.

Training can and should be ongoing (e.g., refresher skills training). One way to keep volunteers' training current is to certify those who have completed training and issue cards with expiration dates.

Step 6: Supervision and Evaluation

Good supervision is essential to volunteer success. A good supervisor:

- Delegates effectively.

- Establishes performance expectations.

- Acts as a coach and team builder.

- Communicates effectively.

- Knows how to listen, and is receptive to information from others.

- Assists staff in developing their skills.

- Gives constructive feedback and takes corrective action, when needed.

- Recognizes staff for their contributions.

Recognition is a critical component of supervision because it is one of the keys to maintaining volunteer interest and, therefore, volunteer retention. Recognition can be formal or informal.

Guidelines for evaluations include:

- Keeping evaluations confidential.

- Making sure comments are fair.

- Focusing on the work, not on the individual.

Step 6: Supervision and Evaluation (Continued)

- Following agency guidelines for disciplinary procedures. (Disciplinary policy should be covered during orientation.) Corrective action may include:

  - Additional training or supervision.
  - Reassignment.
  - Suspension.
  - Termination.

Termination should be reserved for instances when other measures have failed or when there has been gross ethical misconduct. Volunteers should be made aware of grievance procedures to address complaints that cannot be resolved with their supervisors.

Step 7: Evaluating Volunteer Programs

Two-way feedback during the evaluation process allows you to evaluate your volunteer program as well as your volunteers. Evaluating the volunteer program regularly ensures that it is meeting the needs of the:

- Agency.

- Community.

- Volunteer.

Working With VOADs, NGOs, and Other Groups

Some agencies may choose to use the volunteer services of local VOADs and NGOs instead of, or in addition to, maintaining their own volunteer program. The role of a VOAD/NGO Coordinator (as opposed to a Volunteer Program Director) involves:

- Building relationships and acting as a liaison with local community VOADs and NGOs.

- Collaborating with local community VOADs and NGOs to develop and exercise a plan for coordinating volunteer services in an emergency.

The National Voluntary Organizations Active in Disaster (NVOAD) is a consortium of recognized national voluntary organizations whose mission is to foster more effective service to people affected by disaster. VOAD member agencies are coordinated by the State VOAD.

Working With VOADs, NGOs, and Other Groups (Continued)

Working with other agencies requires collaboration on a range of issues, including:

- Shared decisionmaking.

- Sharing information, resources, and tasks.

- Flexibility in dealing with differences among agencies' terminology, experience, priorities, and culture.

- Respect and humility to learn from others' ways of doing things.

Interagency collaboration benefits the community by:

- Eliminating duplication of services.

- Expanding resource availability.

- Enhancing problemsolving.

In addition to VOADs, other sources of volunteers include:

- Nongovernmental organizations (NGOs), such as the Elks and Veterans of Foreign Wars.

- Businesses—especially those with corporate volunteer programs.

- Associations of professionals, such as doctors and nurses.

- The new Freedom Corps.

Keep in mind that volunteers not associated with VOADs may need training in disaster response.

Special Issues

Some of the special issues involved in working with volunteers include:

- Unaffiliated volunteers.

- Safety, risk management, and liability.

- Insurance and workers' compensation.

Unaffiliated Volunteers

People who show up unannounced to volunteer after a disaster present both potential benefits, such as supplementary assistance, and challenges, such as their lack of screening and training.

The preferred way to deal with unaffiliated volunteers is according to a plan developed before a disaster occurs. Such a plan should detail volunteer and donations management-related outreach and education programs, procedures to activate mutual aid such as the Emergency Management Assistance Compact, communications and facilities management, a Volunteer/Donations Coordination Team, a call center, relevant points of contact, safety and security, and demobilization. Plans should also address the important role that emergency public information should play in discouraging and/or channeling unaffiliated volunteers. The media can be enlisted to publicize a phone number that those wishing to volunteer could call for instructions.

Channeling unaffiliated volunteers on site can be done by a Volunteer Coordinator with the use of a skills survey.

Safety, Risk Management, and Liability

The law generally requires that a person act with the care that a reasonable person would exercise to prevent harm. Volunteers for government agencies may be subject to the Federal Tort Claims Act, but are also covered by the Federal Volunteer Protection Act of 1997.

If a volunteer assisting in the disaster response accidentally harms a person or their property, the law generally recognizes that the employer is responsible for the actions and inactions of his or her employee—including volunteers.

Most of the laws limiting the liability of volunteers are State laws.

There are three keys to minimizing susceptibility to lawsuits:

- Train volunteers well.

- Supervise volunteers.

- Document their training and actions.

Insurance and Workers' Compensation

Liability insurance is available to both organizations and individuals. In addition, some States provide Workers' Compensation coverage to volunteers *registered* with an established agency.

Managing Volunteer Stress

Witnessing death, injury, and devastation can cause extreme stress reactions in some volunteers. In addition, working long hours and skipping meals can also contribute to volunteer workers' stress.

Take these steps before, during, and after a disaster to manage stress:

- **Before:** Address stress during volunteer orientation and/or in stress management training seminars.

- **During:** Volunteers should be appropriately matched to their job assignments, get regular meals and breaks, and rotate out at the end of a reasonable-length shift.

- **After:** Invite a trained mental health professional to hold a Critical Incident Stress Debriefing (CISD).

The Final Step

You have now completed IS 244 and should be ready to take the final exam.

Complete the final exam in the back of the book by marking the correct responses.

To submit the final exam online, log onto http://training.fema.gov/EMIWeb/IS/is244.asp and select "Take the Final Exam." After you have selected the final exam link and the online answer sheet is open, transfer your answers, and complete the personal identification data requested.

Appendix A: Job Aids

Job Aid 2.1: Checklist for Determining Volunteer-Staff Climate

Instructions: *Review each of the statements listed below and mark those that you think accurately reflect the climate in your organization. When you finish, review the list. If only a few boxes are checked, you have some work to do to develop a healthy volunteer program.*

☐ Our organization is stable and conflict-free, with a healthy work environment.

☐ Agency policy places high priority on volunteer involvement.

☐ We have established clear, realistic goals for volunteer involvement.

☐ Staff and volunteer roles and responsibilities have been clearly defined and documented.

☐ Volunteer job descriptions were developed with input from staff members and take into account work assistance needs.

☐ The agency recruits and hires staff members who are experienced with and enthusiastic toward working with volunteers.

☐ Our agency ensures that volunteers work primarily with staff members who are receptive to volunteers.

☐ We have established a training program to ensure volunteers and staff members will work together effectively.

☐ Volunteer orientation includes training on sensitivity for staff problems.

☐ We have established a system for rewarding and recognizing volunteers as well as their staff supervisors.

Job Aid 3.1: Basic Criteria for Developing a Volunteer Job

☐ Is there meaningful work for volunteers to do? (Consider its significance to the agency and how to explain the need for the job to potential volunteers.)

Can the work be done by volunteers? (Consider ability to split tasks into part-time work, skill requirements for the job, and whether tasks would be short- or long-term assignments.)

☐ Is it cost-effective to use volunteers? (Consider time, energy, and money for recruitment, orientation, and training of volunteers.)

☐ Is a support framework in place for a volunteer program, including:

- A volunteer manager?
- A volunteer policy?
- Volunteer work space?
- Insurance covering volunteers?

☐ How will paid staff work with volunteers? (Consider experience and receptiveness, as well as role and responsibility definitions for staff and volunteers.)

☐ How will you find volunteers with skills to do the job? (Consider recruitment tactics as well as orientation and training programs.)

☐ Has the agency committed to volunteer involvement with clear policy?

Job Aid 3.2: Sample Application

Name: _____

Address: _____

City: _____ State: _____ Zip: _____

Phone: (Home) _____ (Work)_____

Contact in an emergency: _____ Phone: _____

I. Skills and Interests

Education: Degree _____ Institution _____ Dates attended _____

License(s) held: _____ Language(s) spoken fluently: _____

Hobbies, skills, and interests: _____

Occupation: _____Employer: _____

Address:_____ Phone: _____

II. Experience (paid and volunteer, beginning with the most recent):

Position   Organization   Dates

_____

_____

_____

_____

_____

III. Volunteering Preferences

Is there a particular type of volunteer work in which you are interested? [A checklist of options can be included here.]

_____

Availability (days and hours): _____

Do you have access to a vehicle that you can use for volunteer work? ___ Yes ___ No

How did you hear about our agency? _____

Job Aid 3.2: Sample Application (Continued)

IV. References

Give the names and contact information for three people (not relatives) who know you well and can attest to your character.

_____

_____

_____

V. Verification and Consent for Reference and Background Check

I verify that the above information is accurate to the best of my knowledge.

I give [name of agency] permission to inquire into my educational background, references, licenses, police records, and employment and/or volunteer history. I also give permission to the holder of any such information to release it to [name of agency].

I hold [name of agency] harmless of any liability, criminal or civil, that may arise as a result of the release of this information about me. I also hold harmless any individual or organization that provides information to the above-named agency. I understand that [name of agency] will use this information only as part of its verification of my volunteer application.

_____	_____
Name (please print)	Social Security Number
_____	_____
Signature	Date
_____	_____
Witness	Date

Job Aid 3.3: Volunteer Interview Record

Name of Volunteer_____

Interviewer_____ Date_____

I. Review of Application Form

II. Questions

1. Why do you want to volunteer with our agency? What do you hope to achieve?

3. What kind of work do you most enjoy? Least?

4. Do you work best alone or with others? Why?

5. What kind of supervision do you prefer?

6. What questions do you have?

III. Match with Volunteer Positions

Discuss potential volunteer positions and check match of interest, qualifications, and availability. Ask if there are any physical limitations.

Job Aid 3.3: Volunteer Interview Record (Continued)

To be completed after the interview:

IV. Interviewer Assessment

Appearance:

Disposition/Interpersonal skills:

Reactions to questions:

Physical restrictions:

V. Recommended Action

___ Place as _____

___ Consider/Hold in reserve for the position of _____

___ Investigate further

___ Refer to _____

___ Not suitable for agency at this time

VI. Notification

Volunteer notified of agency decision by (method) _____ on (date)_____

Job Aid 3.4: Orientation Checklist

Before the volunteer(s) arrive(s):

___ Prepare paid staff.

___ Assign a one-on-one mentor.

___ Set up the video presentation and/or confirm date and time with speakers.

___ Collect necessary items (handbook or manual, I.D. tags, etc.).

On arrival:

___ Welcome the volunteer(s).

___ Introduce the volunteer(s) to the staff (paid and volunteer).

___ Review administrative details (phones, parking, restrooms, breaks and lunch, check in/out procedures, etc.)

___ *Optional:* Give a tour of the facility.

Materials you should give volunteers:

___ Mission statement

___ Summary of goals and/or long-range plan

___ Organizational chart

___ Policies and procedures (including emergency procedures)

___ Confidentiality policy

___ *Optional:* Map of facility

What you should tell volunteer(s) about your agency:

___ Mission and goals

___ Background and history

___ Organizational structure

___ Funding base

___ The role of volunteers in the agency

___ The agency's role in the community

___ How the agency relates to other community organizations

Job Aid 3.5: Sample Training Record

Name: _____   Position: _____

Course Title	Course Code	Date Completed	Expiration Date	HR Verification

Job Aid 3.6: Ways to Recognize and Motivate Volunteers

Informal

- Address a volunteer by name.

- Say "thank you."

- Write a thank you note.

- Say "good job."

- Treat a volunteer to coffee.

- Take him or her to lunch.

- Ask how work is going and stop to listen and discuss the response.

- Ask for input.

- Include volunteers in staff meetings.

- Include volunteers in an orientation video.

Formal

- Give annual recognition at an appreciation banquet.

- Hold an awards ceremony during National Volunteer Week.

- Throw a holiday party for volunteers.

- Place a photo and article in the local newspaper featuring volunteers.

- Place a "Volunteer of the Month" photo on the agency bulletin board.

- Present volunteers with plaques, certificates, pins, t-shirts, coffee mugs, etc.

- Ask a volunteer to serve on an advisory board.

- Offer advanced training.

- Give more responsibility, such as the opportunity to train or supervise other volunteers.

Job Aid 4.1: NVOAD National Membership

Adventist Community Services

American Baptist Men

American Radio Relay League, Inc

American Red Cross

Billy Graham Rapid Response Team

Brethren Disaster Ministries

Catholic Charities USA

Christian Reformed World Relief Committee

Churches of Scientology Disaster Response

Church World Service

City Team Ministries

Convoy of Hope

Episcopal Relief and Development

Feeding America

Feed The Children

Foundation of Hope - ACTS World Relief

Habitat for Humanity International

Hands on Disaster Response

Hope Coalition America (Operation Hope)

HOPE worldwide, Ltd.

Humane Society of the United States

International Critical Incident Stress Foundation

International Relief and Development (IRD)

Latter-day Saint Charities

Lutheran Disaster Response

Mennonite Disaster Service

Mercy Medical Airlift

National Association of Jewish Chaplains

National Baptist Convention USA

National Emergency Response Team

National Organization for Victim Assistance

Nazarene Disaster Response

Noah's Wish

Operation Blessing

Points of Light Foundation/Hands On Network

Presbyterian Disaster Response

REACT International, Inc.

Samaritan's Purse

Save The Children

Society of St. Vincent DePaul

Southern Baptist Convention/NAMB

The Salvation Army

Tzu Chi Foundation

United Church of Christ

United Jewish Communities

United Methodist Committee On Relief

United Way of America

World Hope International

World Vision

For more information on National Voluntary Organizations Active in Disaster, visit www.NVOAD.org.

Job Aid 5.1: Skills Survey

Name _____Address _____

Please indicate the areas that apply to you and return this survey to the Volunteer Coordinator.

**PLEASE CHECK ANY OF THE FOLLOWING IN WHICH YOU HAVE EXPERTISE & TRAINING.
CIRCLE YES OR NO, WHERE APPROPRIATE.**

_____ First Aid (current card yes/no)   _____ CPR (current yes/no)   _____ Triage   _____ Firefighting

_____ Construction (electrical, plumbing, carpentry, etc.)   _____Running/Jogging

_____ Emergency Planning   _____ Emergency Management   _____ Search & Rescue

_____ Law Enforcement   _____ Bi/Multi-lingual (what language (s)) _____

_____ Mechanical Ability   _____ Structural Engineering   _____ Bus/Truck Driver
                                                                            (Commercial Driver's License)

_____ Shelter Management   _____ Survival Training & Techniques   _____ Food Preparation

_____ Ham Radio Operator   _____ CB Radio   _____ Journalism

_____ Camping   _____ Waste Disposal   _____ Recreational Leader

DO YOU HAVE EQUIPMENT OR ACCESS TO EQUIPMENT OR MATERIALS THAT COULD BE USED IN AN
EMERGENCY? _____ YES _____ NO

PLEASE LIST EQUIPMENT AND MATERIALS.

_____

_____

COMMENTS _____

_____

Appendix B: Acronym List

AARP	American Association of Retired Persons
BVCs	Business Volunteer Councils
CERT	Community Emergency Response Team
CISD	Critical Incident Stress Debriefing
CPR	Cardio Pulminary Resuscitation
CVCs	Corporate Volunteer Councils
EMI	Emergency Management Institute
EOC	Emergency Operations Center
EOP	Emergency Operations Plan
FEMA	Federal Emergency Management Agency
ICS	Incident Command System
NGO	Nongovernmental Organization
NIMS	National Incident Management System
NRF	National Response Framework
NVOAD	National Volunteer Organizations Active in Disaster
SOPs	Standard Operating Procedures
VFW	Veterans of Foreign Wars
VOADs	Volunteer Organizations Active in Disaster